Biogas - Recent Advances and Integrated Approaches

Edited by Abd El-Fatah Abomohra, Mahdy Elsayed, Zuzeng Qin, Hongbing Ji and Zili Liu

Published in London, United Kingdom

IntechOpen

Supporting open minds since 2005

Biogas - Recent Advances and Integrated Approaches
http://dx.doi.org/10.5772/intechopen.87961
Edited by Abd El-Fatah Abomohra, Mahdy Elsayed, Zuzeng Qin, Hongbing Ji and Zili Liu

Contributors
Yuni Krisyuningsih Krisnandi, Dita Arifa Nurani, Muhammad Reza, Bayu Adi Samodro, Suwardiyanto Suwardiyanto, Nirwan Susianto, Indah Revita Saragi, Russell Howe, Aminah Umar, Sung-Min Choi, Artur Wodołażski, Saeed Alshihri, Hamid Audah Al-Megren, Shohei Riya, Lingyu Meng, Yuexi Wang, Chol Gyu Lee, Sheng Zhou, Koki Toyota, Masaaki Hosomi, Winardi D. Nugraha, Lathifah Laksmi Pradita, Lakshmi Machineni, Gangagni Rao Anupoju, Maria Laura Mastellone, Xu Yang, Salma Mohamed Saleh Omer, Issa Korbag, Hanan Boghazala, Mousay Ahmeedah Aboubakr Abusasiyah, R. Aparna Rao

First published in London, United Kingdom, 2021 by IntechOpen
IntechOpen is the global imprint of INTECHOPEN LIMITED, registered in England and Wales, registration number: 11086078, 5 Princes Gate Court, London, SW7 2QJ, United Kingdom
Printed in Croatia

British Library Cataloguing-in-Publication Data
A catalogue record for this book is available from the British Library

Additional hard and PDF copies can be obtained from orders@intechopen.com

Biogas - Recent Advances and Integrated Approaches
Edited by Abd El-Fatah Abomohra, Mahdy Elsayed, Zuzeng Qin, Hongbing Ji and Zili Liu
p. cm.
Print ISBN 978-1-83962-668-5
Online ISBN 978-1-83962-669-2
eBook (PDF) ISBN 978-1-83962-670-8

For EU product safety concerns:
IN TECH d.o.o., Prolaz Marije Krucifikse Kozulić 3, 51000 Rijeka, Croatia, info@intechopen.com or visit our website at intechopen.com.

Meet the editors

Abd El-Fatah Abomohra, is a Professor of Environmental Engineering at the Department of Environmental Engineering, School of Architecture and Civil Engineering (SACE), Chengdu University, China. He received his Ph.D. in Phycology through a cooperation program between Tanta University, Egypt, and Hamburg University, Germany, funded by Deutscher Akademischer Austauschdienst (DAAD). His major is interdisciplinary microbiology and environmental engineering. He has extensive research experience and finite vision in the field of biofuel production. In 2020, he was nominated among the top 10 publications and citations at Tanta University. His research group is primarily working on bioenergy production from different biomass feedstocks and wastes.

Dr. Mahdy Elsayed received his master's degree in Biosystems Engineering from the Agricultural Engineering Department, Cairo University, Egypt. He obtained a Ph.D. in Environmental and Energy Engineering from the College of Engineering, Huazhong Agricultural University, China. His main research interests include biosystems engineering and biorefineries for cleaner/sustainable production of biofuels and bioproducts. He is currently focusing on the pretreatment and conversion of biowaste as well as other second-generation feedstocks for efficient utilization in biofuel production.

Zuzeng Qin is a professor and doctor's supervisor at the School of Chemistry and Chemical Engineering, Guangxi University, China, where he also obtained a Ph.D. His research interests include heterogeneous catalytic hydrogenation and photocatalytic hydrogen production. He is an editorial board member of *Advanced Composites and Hybrid Materials.*

Hongbing Ji is a professor and doctor's supervisor at Sun Yat-sen University, China. He received a Ph.D. from the South China University of Technology. His research areas include the environmentally friendly catalytic process and green chemical technology. He is director of the Fine Chemical Research Institute, and dean of the Huizhou Research Institute, both at Sun Yat-sen University. He is also an editorial board member of the *Journal of Chemical Technology and Biotechnology* and the *Chinese Journal of Chemical Engineering.*

Zili Liu is a Professor of Chemical Engineering at Guangzhou University. He received a Ph.D. from the South China University of Technology. His research areas include the environmentally friendly catalytic process and green chemical technology. He is also the dean of the Institute of Energy and Catalysis, Guangzhou University, and an editorial board member of the *Journal of Chemical Engineering of Chinese Universities.*

Contents

Preface

The exponential growth of the global population and concurrent fast industrialization has led to massive generation of municipal wastes, with raised challenges of safe disposal. The proper management of municipal wastes through recycling is an essential approach for global sustainable development. So far, many countries have established regulatory guidelines for different waste management routes and pollution control measures. However, most of the applied routes are waste dumping, composting, or direct discharge in water bodies without adequate pretreatment, which seriously threatens the environment and humans. Thus, proper waste segregation and separation provide an efficient option for converting waste into energy. On the other hand, energy demand correlates with population growth, as population directly increases eventual energy demand. Thus, global energy demand and environmental pollution are two inevitable issues that demand the discovery and development of alternative energy sources. Waste-to-energy is a widely used statement for efficient waste management, which is getting much attention. For almost two decades, biofuel production from biowastes receives paramount of importance and eventual growth and development. In general, it is widely accepted that biowaste-derived fuels can reduce the current dependence on fossil-based products. Among different biofuel production routes, anaerobic digestion (AD) is by far the single most important technology for providing clean renewable biogas to millions of people in rural areas of developing countries. AD technology has several inherent benefits ranging from generating renewable energy and remediating biowaste curtailing CO_2/CH_4 emissions to improving health/hygiene and overall socio-economic status of rural communities in developing nations.

This book provides basic knowledge and recent research on biogas production, focusing on the enhancement of biomethane and integrated production routes. It comprises nine chapters divided into two sections. The first section focuses on the production and upgrading of biogas, and the second section presents some integrated approaches of biogas production with other applications such as microalgae cultivation and biofertilizer production.

We would like to express our gratitude to the faculty members of the participating institutes. We are also grateful to IntechOpen's Commissioning Editors and Author Service Managers for being generously helpful throughout the process of creating and publishing this book.

Abd El-Fatah Abomohra
Professor,
Department of Environmental Engineering,
School of Architecture and Civil Engineering,
Chengdu University,
China

Dr. Mahdy Elsayed
Faculty of Agriculture,
Department of Agricultural Engineering,
Cairo University,
Giza, Egypt

Zuzeng Qin
Professor,
School of Chemistry and Chemical Engineering,
Guangxi University,
Nanning, P. R. China

Hongbing Ji
Professor,
School of Chemistry,
Sun Yat-sen University,
Guangzhou, P. R. China

Zili Liu
Professor,
School of Chemistry and Chemical Engineering,
Guangzhou University,
Guangzhou, P. R. China

Section 1

Biogas Production and Upgrading

Chapter 1

Recent Advances of Biogas Production and Future Perspective

Issa Korbag, Salma Mohamed Saleh Omer,
Hanan Boghazala
and Mousay Ahmeedah Aboubakr Abusasiyah

Abstract

The production of biogas via anaerobic digestion (AD) provides significant benefits over other techniques of bioenergy production. Biogas consists of several undesired components, such as H_2S, CO_2, nitrogen, hydrogen, oxygen, and water vapor, which contribute to lower the calorific value when compared with natural gas. The pollutants founded in low concentration effects the biogas commercial application in large scale, and therefore it must be taken out before usage. Various cleaning and upgrading techniques to improve the quality of raw biogas are discussed and categorized into physiochemical and biological technologies. Advanced techniques, such as hydrate separation, cryogenic separation, biological methods, membrane enrichment, in-situ upgrading, multistage and high-pressurized anaerobic digestion, represent the modern developments in biogas upgrading techniques. Biogas is a renewable green source of energy, and presently, it is utilized in a lot of developing countries as an alternative and energy renewable source for a broad range of applications. Most countries are in the process of instituting legislation to regulate the biogas industry. Biogas is considered to be the future of renewable and sustainable energy.

Keywords: biomass, biogas, biofuels, bioenergy, renewable energy

1. Introduction

The demand of energy has been increased over the years as the sequence by increasing of the world population [1]. Fossil fuels are exhausting and the emission products of these fuels have been causing some damages to the environment. The scientists in the world are focusing on developing alternative methods of energy production [2]. Bioenergy is an energy obtained from any fuel that is originated from biomass, which includes recently living organisms and their metabolic by-products [3]. Biomass is defined as all animal and plant material on the Earth's surface. Hence, collecting biomass, such as manure, crops, or trees, and employing it to produce electric power, heat, or motion is bioenergy [4, 5]. If not managed optimally, the large amounts of biomass, livestock's manure, agro-industrial waste, and slurries produced today as well as the wet organic waste streams represent a constant pollution risk with a potential negative impact on the environment [6]. Biofuels are defined as fuels made from biomass resources, or their processing and

IntechOpen

conversion derivatives [3, 5]. Biofuels are eco-friendly and renewable resources of energy and hence have been receiving attention as an alternative energy source [2]. The organic part of nearly any form of biomass, involving industrial effluents, sewage sludge, and animal waste, can be decomposed via AD into carbon dioxide and methane mixture called as biogas and is considered an alternative green energy resource. Methane (CH_4) is the most important component of biogas because it has the highest energy density among the biogas components. Therefore, the high CH_4 content of biogas is desired [7, 8]. Biogas was first identified 600 years ago as originating from decomposing organic matter. More recently, in 1884, Louis Pasteur investigated it sourced from animal waste, suggesting it as an appropriate fuel for the lighting of street lamps. Biogas primarily consists of methane (CH_4), in a range of 50–75%, and carbon dioxide (CO_2), at 25–50%, with minor amounts of other compounds, such as hydrogen (0–1%), nitrogen (0–10%), which could originate from air saturated in the influent, vapor water (H_2O) at concentrations of 5–10%, or higher at thermophilic temperatures, derived from medium evaporation, hydrogen sulfide (0–3%), which is produced from reduction of sulfate contained in some waste-streams, ammonia (NH_3) originating from hydrolysis of proteinaceous materials or urine and oxygen (0–2%), which is entering the process from the influent substrate or leakages, hydrocarbons at concentrations of 0–200 mg/m^{-3}, trace carbon monoxide (CO), and siloxanes at concentrations of 0–41 $mg\ m^{-3}$, originating for example from effluents from cosmetic medical industries. The relative content of CH_4 and CO_2 in biogas is mainly dependent on the nature of the substrate and pH of the reactor [7–19]. Typical components and impurities influence the quantity and quality of the biogas. CO_2 and N_2 lowers the calorific value, CO_2 also causes corrosion and damages to alkali fuel cells. H_2S spoils catalysts, causes excessive corrosion and deterioration of lubrication oil, generates harmful environmental emissions, and corrodes the engines of biogas purification machinery. N_2 and NH_3 increase the anti-knock properties of engines and NH_3 also damages fuel cells. Water vapor causes corrosion of equipment and piping system leading to damage instruments and plants. Siloxanes acts like an abrasive and damages engines. Dust blocks nozzle and fuel cell [18–23].

Biogas is flammable, smokeless, hygienic, colorless, odorless, and has bad eggs odor whether not desulfurized. It has an energy content of 37.3 MJ/m^3, explosion limits 6–12% biogas in air, ignition temperature 650–750°C, specific gravity 0.847–1.004, and calorific value 4740–7500 $kcal/Nm^3$ [18, 24–28]. Biogas is an environmentally friendly, a renewable, clean, cheap, high quality, and versatile fuel which is generated in digesters filled with the feedstock. It is considered an alternative green energy resource. It can be utilized for different energy services like heat, combined heat and power, or a car fuel [7, 8, 29].

Biogas technology is used to convert the organic waste into energy. The use of energy and manure can lead to social economic benefits, green environment, and also contributes towards sustainable development [30–32]. Biogas technology is also a source of nutrient-rich organic fertilizer and the effluent slurry produced as a result of biogas technology is also helpful for algae growth, fish production, and seed germination [24]. Biogas technique is applied to small-scale and large-scale uses involving electric power production. It is a mixture of gases of which the composition relies on substrates and AD process conditions like retention time, temperature, and pH. Biogas is one of the main products of the AD of organic substances.

Anaerobic digestion (AD) is considered as a biological process that degrades organic substances by the actions of microbial communities in the absence of oxygen. In fact, AD can be divided into four stages, as seen in **Figure 1**, which are hydrolysis, acidogenesis and this stage is considered as acid-producing, acetogenesis

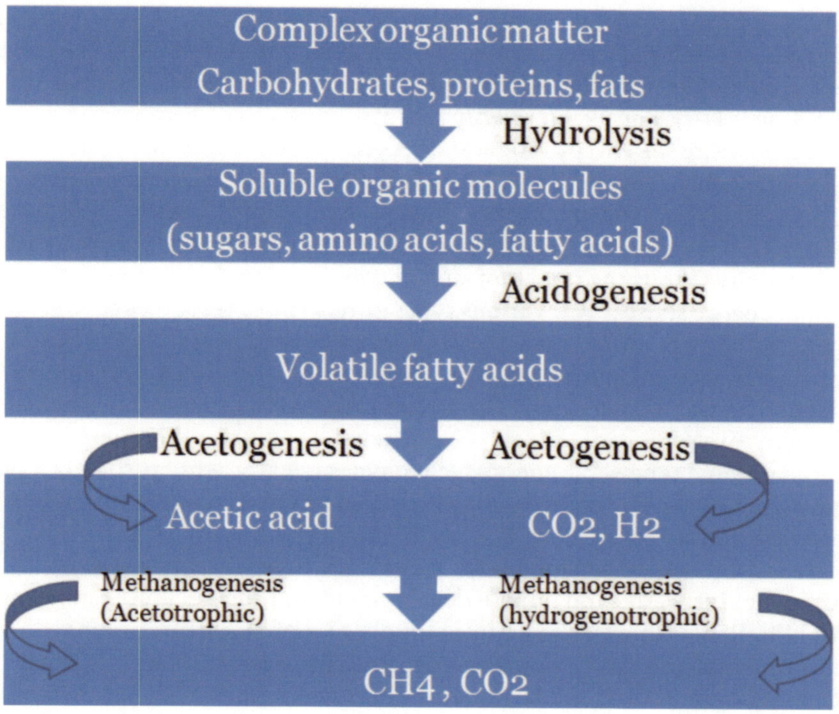

Figure 1.
The steps involved in anaerobic digestion.

and this phase is also called acetic acid-producing, and methanogenesis and this final step is known as methane-producing [8]. It is quite important to purify raw biogas and upgrade it to a high-quality fuel standard, in order to increase the calorific value and reduce undesired components, such as H_2S and CO_2, which are damaging the utilization systems. This process is well known as biogas cleaning and upgrading [33, 34]. Biogas could be simply upgraded into biomethane or renewable natural gas (RNG), which is similar to natural gas that produced from nonrenewable fuel sources. It contains about 90% or greater of methane. RNG could be replaced for natural gas and could be used as fuel for cars that can run on natural gas and to provide gas to natural gas grid. Upgrading of biogas to biomethane is considered as one of the technologies that has got a lot of attention in the bioenergy industry [8, 35]. Biogas could play a key role in the developing market for renewable energy and the utilization of biogas in the world is expected to be doubled in the next years, ranging from 14.5 GW in 2012 to 29.5 GW in 2022 [20, 36, 37].

2. Renewable energy resources

The term energy can be generally defined as the amount of force or power when applied can move one object from one position to another, or it defines the capacity of a system to do the work. The most important characteristic of energy is the possibility to convert one of its forms to another. Generally, the energy technologies are the man-made devices, equipment, and systems used to capture, convert, store, and transport energy from the energy resources [38]. Energy is an important demand in our daily life as a way of enhancing human development leading to productivity and economic growth. Energy is a key driver for agriculture, industries, and service sectors that influence economic development [39]. The term energy sources refer to the output forms of energy from the man-made energy technologies, while the energy resources refer to the naturally available forms of energy [38]. The resources of energy are divided into three groups: nuclear resources, renewable resources, and fossil fuels [40]. Renewable energy is called "renewable" because the sources harnessed to create the energy renew and replenish themselves constantly and within a reasonably short period of time (i.e., months or years, not centuries) [41]. Thus, renewable energy sources renew themselves naturally without being drained in the earth [39, 41]. Renewable energy techniques give a great opportunity for reduction of greenhouse gas emission and decreasing global warming via replacing traditional energy sources [6]. Various types of renewable resources such as hydro-thermal, geothermal, solar, wind, ocean (tide and wave), heat from the Earth's interior, and biogenic (biomass) energies are available and they give the possibility to produce consistent power [38]. These renewable energy sources are also often called alternative sources of energy [42]. Alternative or renewable energy sources include traditional renewable energy technologies (such as wind turbines) as well as innovative new technologies, such as hydrogen internal combustion engines, and hydrogen-based fuel cells [38].

There is a direct relationship between renewable energy and sustainable development via its effect on human development and sustainable economic growth [43]. Renewable energy sources supply a lot of opportunities in reduction of environmental and health impacts, energy access, energy security, climate change mitigation, social, and economic development [6, 38, 44–46]. Renewable energy resources will play an important role in the world's future [6, 40]. The technologies developed to exploit renewable energies are known as renewable energy technologies (RET) or clean technologies or green energy. Sustainable economic and industrial growth also requires safe and sustainable energy resources [47–49]. Maximizing resource (material and energy) recovery and minimizing environmental impacts such as contribution to the global warming are important objectives in the solid waste management (SWM) sector, which is considerably developed over the past century [50, 51].

3. Sources of biogas

There are a variety of wastes that can be used as sources, or feedstocks, that are fed into the digester to produce biogas. Most of biodegradable organic compounds could be transformed into biogas by anaerobic digestion, and biodegradability is the characteristic indicating to what extent this is possible. Raw materials for producing biogas by anaerobic digestion are biomass feedstocks which include; municipal solid waste (MSW), industrial solid wastes and industrial wastewaters, food waste, livestock manure, sewage sludge, agricultural manures, catch crops, energy crops, and micro-algae [12, 52–57]. The largest resource is represented by animal manure and slurries from cattle and pig production units as well as from poultry, fish, fur, etc. [58–61]. Manure of animal is considered as a main carbon source for biogas and involves pig manure, cattle, and poultry. The total solids present in animal manures consist of 90% moisture content and volatile solids. It performs as perfect substrate because of its great buffering capacity [62–64]. In most countries, sewage sludge and agricultural manures have been the principal sources for some time, complemented by slaughter-house, dairy, and restaurant waste. More recently, biogas plants are increasingly using municipal solid waste (MSW), industrial solid wastes, and industrial wastewaters as feedstocks [12].

Nowadays, sludge from municipal wastewater treatment plants is considered as major source of organic matter for biogas generation in Sweden. Other familiar substrates for biogas generation in co-digestion plants involve source-sorted food waste and manure, slaughterhouse waste, and waste from the feed and food industries [65]. Sewage sludge is usually used as a feedstock to provide energy to power sewage treatment works. For many years, sewage sludge and agricultural manures was the principal inputs, making up over 80% of the total.

However, more lately, the manufacturers have been testing with biogas particular agricultural crops, involving maize and rapeseed. Both the crop itself and the generated fodder (silage) are utilized. Animal waste is increasingly used as a feedstock throughout the world. In the EU, there are now over 750 biogas plants processing animal waste, many of them on a large scale. Organic waste from households and municipal authorities is also an important source of biogas [12].

The urban solid waste production, or municipal waste, increases with population growth, high economic activity, and goods production. Biogas has the potential to be produced from widely available, abundant raw materials, including agricultural residues (e.g., animal manure), landfill and food waste, and aquatic biomass and lignocellulosic raw materials [66, 67]. Wood, agricultural residues, and dung of animal are the sources of energy for biogas technique [24, 68]. The use of wastewater from inorganic sources, such as chromium, has also been studied as an alternative for energy production, a more environmentally sustainable approach that avoids landfill disposal of these wastes [69, 70]. Industrial waste and wastewater have potential uses in biogas production due to their characteristics, such as high organic load [66, 71]. Algae are considered as a potential biomass feedstock for decreasing our dependence on nonrenewable energy sources for electric power, transportation, and heat production [53].

Livestock manure, i.e., dung of cow is an efficient feedstock for biogas production getting high cumulative biogas yield with steady performance, with a continuous process. Therefore, cow dung is more favorable in the biogas process [72, 73]. In general, there are different kinds of biomass resources to produce biogas, including animal manures, municipal solid wastes, food wastes, industrial wastes, agricultural residues, poultry wastes, forestry wastes, microalgae, and some dedicated energy crops [54–57].

4. Biogas production processes

Generation of biogas gives a multiuse carrier of renewable energy, as methane can be utilized for substituting of nonrenewable source of fuels in both heat and electricity production and as a car fuel. AD of wastes, energy crops, and residues is of growing interest in order to decrease the greenhouse gas emissions and to promote a sustainable development of energy supply [74]. Anaerobic digestion is a technology with proven efficiency, being widely used in the stabilization of industrial wastewater, urban solid waste, animal manure, and sewage sludge [66]. There are many benefits associated with anaerobic digestion technology, which include mass reduction, odor removal, pathogen reduction, less energy use, and more significantly, the energy recovery in the form of methane [75, 76]. The aim of anaerobic digestion process is the production of a methane-rich biogas through biological decomposition of organic matter, in an oxygen-free environment. An aerobic digestion is considered as a low-cost an eco-friendly waste management process, thus it reduces the emission of greenhouse gases. In the meantime, it stabilizes and reduces the wastes. One of the major advantages of an aerobic digestion is its adaptability to deal with a wide range of organic substrates. The produced biogas can be used for power and heat production, or can be upgraded and used as vehicle fuel in the transport sector. In addition, the by-product of AD, the "digestate residue," can be further utilized as a fertilizer on the agricultural land [50]. There are different process types which can be applied for biogas generation, which are classified in dry and wet fermentation systems [21, 77, 78]. Wet digester systems are constantly applied using vertical stirred-tank digester with various stirrer kinds dependent on the source of the feedstock.

Biomass is utilized as substrates for biogas generation as long as it consists of hemicelluloses, cellulose, carbohydrates, proteins, and fats as major constituents. Only powerful lignified organic materials, e.g., wood, are not suitable due to the slowly anaerobic decomposition. The composition of biogas and the methane yield depends on the feedstock type, the digestion system, and the retention time [78]. Maximal gas yields and theoretical methane contents of substrates for biogas production are carbohydrates 790–800 biogas (Nm^3/t TS), 50% CH_4 and 50% CO_2, carbohydrates only in the form polymers from hexoses, not inulins and single hexoses, raw protein 700 biogas (Nm^3/t TS), (70–71)% CH_4 and (29–30)% CO_2, and finally raw fat 1200–1250 biogas (Nm^3/t TS), (67–68)% CH_4 and (32–33)% CO_2 [79].

4.1 Biochemical process

Anaerobic digestion involves bacterial fermentation of organic wastes in the absence of free oxygen. Methane fermentation is a complex process, the fermentation leads to the breakdown of complex biodegradable organics in a four-stage process: hydrolysis, acidogenesis, acetogenesis, and methanogenesis [50, 74, 80].

First stage (hydrolysis process): large protein macromolecules, fats, and carbohydrate polymers (such as cellulose and starch) are broken down through hydrolysis to amino acids, long-chain fatty acids, and sugars.

Second stage (acidogenesis process): the products obtained in first step are then fermented via acidogenesis to form volatile fatty acids, valeric acid, propionic, principally lactic, and butyric. Third stage (acetogenesis): bacteria devour these fermentation products and produce acetic acid, hydrogen, and carbon dioxide. Fourth stage (methanogenic): organisms feed on the hydrogen, acetate, and a few of the carbon dioxide to generate methane [81]. Three biochemical pathways are used by methanogens to achieve this:

$$4CH_3COOH \rightarrow 4CO_2 + 4CH_4 \text{ (acetotrophic pathway)} \quad (1)$$

$$CO_2 + 4H_2 \rightarrow CH_4 + 2H_2O \text{ (hydrogenotrophic pathway)} \quad (2)$$

$$4CH_3OH + 6H_2 \rightarrow 3CH_4 + 2H_2O \text{ (methylotrophic pathway)} \quad (3)$$

Biogas is a multipurpose renewable green energy source, which can be simply used to substitute nonrenewable energy source, in heat and power generation, and as gaseous car fuel. Biomethane can also substitute natural gas as a feedstock for producing chemical materials. The biogas generation during AD provides vital benefits over other bioenergy generation technologies. It is admitted as one of the most energy-efficient and environmentally beneficial technology for generation of bioenergy [82, 83]. Anaerobic digestion is a broadly used technology that provides some benefits over other biofuels generation ways, such as, sustainable biogas production, option for using wastewater and sea water, lower operational costs, maximum biomass utilization, minimum sludge production, lesser energy consumption, and feasibility to recycle nutrients [54, 84, 85].

AD of animal manure provides some socio-economic, environmental, and agricultural benefits via inactivation of pathogens, improved fertilizer quality of manure, and considerable reduction of odors, and last but not least production of biogas generation, as green renewable fuel, for multiple utilizations [58]. The slurry or digestate from the reactor is affluent in ammonium and other nutrients utilized as an organic fertilizer [86, 87]. The European renewable energy directive has set a target to substitute 27–30% of the total energy consumption with renewable energy sources by 2030. It is expected that 14–26% of this renewable energy target could be achieved by biogas from farming and forestry residues [61, 88]. Biogas is presently produced and utilized in Europe. In 2007, Germany was the largest biogas producer in Europe mainly from energy crops, while the UK was the second producer of biogas mainly from landfill sources [50].

There are three common technologies used (in **Figure 2**) to convert biomass to green sustainable products. Thermal approaches that are commonly used to convert biomass into an alternative fuel are: gasification, liquefaction, pyrolysis, and charcoal, while there are two biological approaches that are commonly used to convert biomass into bioenergy: fermentation and anaerobic digestion, as shown in **Figure 2**. This research is going to focus on an anaerobic digestion to produce biogas.

The anaerobic co-digestion is a choice to settle the drawbacks of single substrate digestion system, being the properties of the substrates and chemical composition, the operating parameters (pH, charge rate, temperature, etc.), the bioavailability, biodegradability, and bioaccessibility, significant parameters to be optimized.

Some of raw materials need to be treated to improve the biogas production. In the past, AD was mostly referred to a single substrate/single output process but recently, co-digestion has become a standard technology in agricultural biogas production in many countries [50, 66]. The anaerobic co-digestion is the simultaneous digestion of more than one substrate with complementary characteristics and has become popular as the digestion of several materials can give higher methane yields than those expected when single materials are treated individually [91–93]. Several of the reasons related with the improvement are associated to the combinations of substrates that result in a positive interaction within the system, reducing negative influences of toxic or inhibitory compounds, affecting C/N ratio and reactor stability, supplementing nutrients, and balancing buffer capacity. Additional benefits of using co-digestion techniques including improved balance nutrients, synergistic effect of microorganism, increased load of biodegradable organic matter, and higher biogas yield [50, 82, 94, 95].

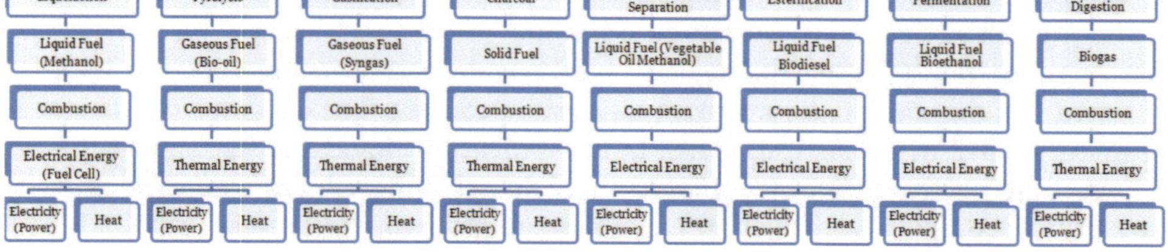

Figure 2.
A schematic of various biomass conversion technologies [5, 89, 90].

4.2 Factors affect biogas production

Production of biogas involves a series of four complex biochemical processes (hydrolysis, acidogenesis, acetogenesis, and methanogenesis) depending upon different factors such as type of substrates, substrate particle size, temperature range, pH, carbon/nitrogen (C/N) ratio, and inoculums concentration [62, 96–99]. There are various factors affecting biogas production from anaerobic digestion. Some of the key factors have been elaborated in detail below.

4.2.1 Hydrogen-ion (pH) concentration control

The hydrogen-ion (pH) concentration in the digesting material affects the anaerobic digestion process. The hydrogen-ion concentration of the culture medium has an immediate effect on microbial growth due to the digestion is prevented by surplus acidity [100, 101]. Methanogens grow better under neutral and a bit alkaline environments. They are died by acidic conditions. Upon stabilization of the aerobic digestion process, the optimum values of pH in the system will be in the range of 7–8.5 with values near to 7 for optimal activity [21, 96, 102–105].

4.2.2 Temperature control

The temperature of MSW influences the success of the digestion process, as the activities of the anaerobes causing waste decomposition are temperature dependent. Optimum performance of an aerobic digestion system is affected largely by the operating temperatures of the reactor. There are three general ranges of temperature each favoring a specific type of microorganisms including; psychrophilic: about 10–20°C, or less than (<30°C), mesophilic: about (30–40°C), and thermophilic: about (50–55°C), or may reach 60°C [50, 106]. Previous studies have shown that anaerobic bacteria exhibit the highest activity within the mesophilic and thermophilic ranges [107]. Extreme cases of either very high or very low temperatures kill the anaerobes, hence inhibiting the whole AD process [96]. The rate of decomposition and gas production is sensitive to temperature, and, in general, the process becomes more rapid at high temperatures [3, 102]. The optimum temperature is 35°C [101, 108]. There are a couple of factors, which contribute to heat generation or transfer in a digester including process reaction, mixing (impellers), as well as heat exchangers (hot water or steam) [50].

4.2.3 Feedstock composition and nutrients

A variety of digester kinds exists for the anaerobic treatment of organic wastes.

Some diversity of biomass feedstocks could be used by anaerobic digestion techniques such as biowaste, agricultural crops, human waste, municipal sewerage and animal manure among others. The quality and quantity of the biogas yield is determined via nature of the feedstock used. Furthermore to the biogas yield, biomass generates vital nutrients and carbon that promote the sustainable growth of the microbes [96, 107, 109]. The selected kind rely on operational factors, involving the nature of the waste to be treated, e.g., its solid content. The Oregon Department of Energy [110], in its classification of kinds of digester, elucidates that "a plug-flow digesters are appropriate for ruminant animal dung having solid concentrations of 11–13%"; "a complete-mix digester is appropriate for manure that is 2–10% solids"; and "a covered lagoon digester" is used for liquid manure of less than 2% solids. The amount and kind of solid contents of the waste they considered were such that the wastes are able to flow on their own or forming slurries with water and finally flowing, and thus can be used in a continuous operation [102].

4.2.4 Carbon/nitrogen (C/N) ratio

The concentrations of carbon and nitrogen determine the anaerobic digestion performance. Anaerobic digestion ideally occurs at C/N ratio ranges between 20:1 and 30:1. For the optimal operation, the ratio of the carbon, to, nitrogen should be about 30:1 in the raw material. Methanogenic bacteria use nitrogen to meet their protein requirements. Whereas, carbon constitutes the energy source for the microorganisms, nitrogen serves to enhance microbial growth. If the amount of nitrogen is limiting, microbial populations will remain small and it will take longer to decompose the available carbon [102]. Consequently, in cases of high C/N ratios higher than the optimum ranges, the nitrogen will be promptly consumed by the bacteria and thus will not react on the excess carbon in the feedstock, hence decreasing the biogas yield. For cases of lower ratios than the limited range, the excess nitrogen will result into ammonia (a strong base) formation, thereby increasing the working pH over the required 8.5 inhibiting the microbes and finally dropping gas generation rates [96, 107]. It has been found that the bacteria in the digestion process use up the carbon present 30–35 times faster than the rate at which they convert nitrogen [102]. The high amount of nitrogen content in animal manures reduces its utilization in anaerobic digestion for biogas generation because of its C/N ratio [62, 111]. To solve this problem, nitrogen-free raw material or carbohydrate-rich source is used to raise the carbon content in the animal manure before the AD process proceedings [62].

4.2.5 Substrate particle size

Pretreatment of biomass should be processed to reduce the particle size and then followed co-digestion to increase the biogas production [62, 112]. The substrate for anaerobic digestion has to be composed of digestible particle sizes. Smaller particles raise surface area for the microbial action of the methanogens as a result growing the biodegradability of the feedstock, hence raising the rate of biogas generation and vice versa for great particles which may clog the digester [96, 113, 114].

4.3 Pretreatment technologies used in biogas production

A number of various treatment processes presently exist for the organic waste management; some more technologically modernized than the others, and some more founded in some countries where the legislation and policy promotes for certain environmental goals. The four alternative systems are currently applied worldwide, i.e., Landfilling, Aerobic Composting, Incineration and Anaerobic Digestion (AD). Anaerobic digestion of organic waste is the most desirable management method and this research is going to discuss it in detail [50, 115, 116].

Organic waste is considered extremely heterogeneous, whereas its moisture content as well as level of impurities differs significantly. Hence, pretreatment before an aerobic digestion is a main process. Diverse types of pretreatment technologies have developed and are successfully installed in many anaerobic digestion plants in the worldwide. The organic waste pretreatment is considered as the major process step in biogas generation plants ensuring flexibility to treat different types of organic waste, efficient extraction of contaminants, high availability of AD plants, substrate homogenization, wear resistance, high biogas yields, energy efficiency, and production of high-quality fertilizers. Organic waste almost includes contaminations, such as glass, metals, stones, and sand; additional systems are also required to deal with such heavy contaminants of the waste. Thus, pretreatment techniques, may be mechanical (e.g., milling), chemical (e.g., acid or alkali treatment), or thermal methods (e.g., steam explosion), are usually applied. Novel pretreatment methods are emerging, which focus on ionic liquid or supercritical CO_2 to solubilize and collect lignin, both increasing biogas production, while also providing additional revenue through lignin collection. Regardless of the pretreatment technique used, this step is an essential consideration for improved biogas production from lignocellulosic feedstocks [50]. In any AD application treating organic waste, a mechanical pretreatment is installed.

A critical unit in industrial biogas plants is mechanical pretreatment; it possibly contains pulpers and shredders. These apparatus are utilized to improve the surface area of tough solid substrates [such as municipal solid wastes (MSW), cardboard, mixed industrial wastes, bulky refuse, waste tires, waste wood and waste papers, etc.] via crushing and breaking down, leading to their more efficient digestion and improved AD process. Different pretreatment technologies are available to reduce the size of the organic waste and to separate the plastic and packaging material from the biodegradable fraction of the waste. Thereby, plants are generally highly flexible to treat all kinds of organic waste without any quality restrictions. Sewage sludge or agricultural biomass, e.g., straw are difficult to degrade an aerobically due to their rigid structure. Therefore, organic waste thermal treatment at high pressure and temperature values is more familiar when treating such types of organic biomass. In contrast, food waste can be efficiently converted in anaerobic digestion systems to biogas by mechanical processing. An efficient pretreatment of organic waste also ensures the production of high-quality fertilizers and hence, the recycling of valuable nutrients back into the natural cycle is achieved and additional, expensive digestate processing after AD can be avoided. High biogas yields in anaerobic digesters are achieved, whether the biodegradable organic material is well crushed in the pretreatment and a large surface are for microbial degradation is achieved [50].

The composition of organic waste come from different areas (commercial, municipal, industrial) differs significantly. The most critical criteria for the selection of an appropriate pretreatment technology are waste composition. Moreover, for the selection of the most appropriate pretreatment system, it is significant to know which kind of AD, i.e., dry or wet digestion systems, should be used to treat the organic waste. Wet anaerobic systems use pretreatment technology to take out the undesirable pollutants before the anaerobic digestion process and are operated at a lower solid concentration. The digestate after anaerobic digestion may be used immediately as high-quality fertilizer and no further digestate treatment (compost refining, post-composting, etc.) is usually required. Amount of biogas generation is high due to the efficient organic waste pretreatment. The preferred technologies to treat wet organic waste such as food leftovers, food waste, packaged food, and organic fraction of MSW are Wet anaerobic digestion systems. Dry anaerobic systems use simpler pretreatment technology before the anaerobic digestion process and are operated at higher solid concentrations. As the efficiency of impurity separation is not enough to utilize the digestate immediately as high-quality fertilizer, further digestate treatment (i.e., compost refining, post-composting) is normally necessary to know whether the input substances are polluted.

Dry anaerobic technology are mostly utilized to know whether the organic waste involves a high percentage of garden waste and also after a mechanical extraction process can be used to treat the organic fraction of MSW [50]. Dry anaerobic digesters are higher solids loading and biomass retention, controlled feeding and spatial niches, pretreatment is simpler, but it has complex and expensive transport and handling of waste, material handling and mixing is difficult, and only structured material can be used [117]. For processing dilute organic slurry with a total solid content of maximum 10–15%, the wet systems are designed. Substrates consist of total solid higher than 15% will be co-digested with co-substrates of lower total solid content, or usually diluted with recirculated or fresh process water. Various ranges of low solid substances have been successfully treated by wet AD technology, involving food industrial effluents and sewage sludge. In contrast, in solid-state fermentation processes, also called dry digestion, the substrates used have high solid content (25–40% TS), thus an essentially different technical approach regarding the waste handling and treatment is needed [50, 118]. Due to the high viscosity in the dry digestion systems, heat and nutrient transfer is not as efficient as it is in wet processes, therefore mixing is very important to prevent local overloading and acidification [119]. In spite of that, conventional mechanical mixers are not appropriate for solid-state processes; instead, recirculation of the waste or re-injection of the produced biogas is often used in these types of reactors to solve the mixing problems [50, 120]. The main benefits of wet anaerobic digesters is dilution of inhibitors with fresh water, but it has some drawbacks including scum formation during crop digestion, high consumption of water and energy, short-circuiting, and sensitive to shock loads [117].

5. Operational methods and reactor designs

Digesters established in worldwide differ in their costs, construction materials, and design complexity [121, 122]. In order to design any anaerobic digester, we need to solve three principal requirements such as: to produce a high volume of high-quality biogas; able to continuously handle a high organic loading rate; and to have a short hydraulic retention time in order to have smaller reactor volume. There are various types of digesters, which are mostly used in the industry involving multistage systems, batch, continuous one-stage system, or continuous two-stage. Further configurations, such as the plug-flow systems, anaerobic sequencing batch reactor (ASBR), tubular reactor, baffled digesters (ABR), upflow anaerobic sludge blanket (UASB) reactor, and anaerobic filters, are also present [50, 104, 123]. Normally, the selection of suitable digester kind is relying on the properties of the main feedstock used, specifically total solid. Feedstocks with high amount of total solid and slurry are generally treated in CSTRs; whereas, soluble organic wastes are mainly digested in upflow anaerobic sludge blanket (UASB) reactors, anaerobic filters, and fluidized bed reactors [124]. Co-digestion is principally implemented in wet single step processes (e.g., CSTR). Continuous systems are fed continuously, while the digestate residue is discharged at the same rate, allowing a steady state to occur, leading to a constant gas production rate. In spite of that, this kind of operation is only possible for substrates, which can be pumped for continuous feeding. Otherwise, a semi-continuous process is applied with a discrete amount of feed several times a day [50]. The main advantages of continuous systems are simplicity in design, operation, and have low capital costs, but they have disadvantages including rapid acidification and larger volatile fatty acids (VFA) production [117].

5.1 Continuous flow stirred-tank reactors (CSTRs)

Continuous flow stirred-tank reactors (CSTRs) are one of high rate digesters and probably the most generally used reactor configurations in biogas generation. They are interesting because of the simplicity of their design compared to other types of biogas digesters. Normally, CSTRs are usually utilized to process slurries with total solids content of 5–10% [50, 125]. Slurries of animal manure and organic industrial wastes are treated using CSTRs. As a disadvantage, CSTRs have long retention times [126] and may be more energy intensive than some of the other types of reactors. Performance of CSTRs is improved by recycling microbial solids, or enhancing retention of the active biomass [50].

5.2 Anaerobic plug-flow reactors (APFRs)

Anaerobic plug-flow reactors (APFRs) are generally long rectangular channels, with the flow entering one end and leaving at the distant end. There is roughly seldom mixing in the flow direction. The channels, or tanks, are mostly placed above ground. Both thermophilic and mesophilic operations are utilized [50, 127]. APFRs are considered one of high rate digester and commercially used for treating different types of organic wastes involving slurries of animal manure, distillery wastewater, and the organic fraction of municipal solid waste [128, 129]. Compared to a single-stage CSTR, plug-flow reactors are mostly more efficient in converting the substrate to biogas and are more stable to operate [50, 130].

The two generally utilized reactor kinds are: continuous stirred-tank reactors (CSTR, using biogas recirculation for mixing or mechanical agitation or effluent), and plug-flow reactors (PFR, where the reactor content is shoved along a horizontal reactor). In dry digestion processes, PFRs are usually utilized to treat substrates with high solid content [131], whereas CSTRs are applied in wet digestion systems. The choice of wet or dry digestion technology relies on the total solid content (TS) of the material treated [50]. Recovery of biogas from manure is widely applied with CSTR and PFR systems in developed countries, likewise covered lagoons, and other kinds of anaerobic reactors are also used [74].

5.3 Anaerobic contact reactor (ACR)

Anaerobic contact reactor (ACR) is consistently a fully mixed mechanically stirred tank with recycle of sludge. The effluent from the tank flows into some kind of a solid-liquid separator (e.g., gravity sedimentation tank, sludge flotation device, lamella clarifier) and the recovered solids are returned to the anaerobic digester. ACRs are efficient of treating high-strength waste with a high concentration of digestible solids due to high concentration of active microbial biomass [132–134]. Hydraulic retention times are short and fluctuations in organic loading are well tolerated. The ACRs are relatively less affected to souring and other inhibitors [128, 135, 136]. Stirred digesters coupled to some type of membrane-based cell retention have proved highly effective in biogas production [50, 137, 138].

5.4 Biofilms

Biofilms are microbial consortia attached to a support material. The support surface is usually inert and may be fixed or suspended. Anaerobic microbial biofilms can effectively digest organic material to produce biogas [139]. A huge mass of immobilized biofilm and mass-transfer upgrading motion of liquid around the film let biofilm reactors to hold high organics loading and bear well any fluctuations in hydraulic or organics loads. Once the biofilm has produced, start-up periods are short compared to the other traditional anaerobic treatment systems [139, 140]. The support material nature affects the improvement of the biofilm and its intensity of attachment, or mechanical steadiness [50, 141].

5.5 Batch reactors

Batch reactors are quick, require inexpensive equipment, and are the simplest to operate since they are fed with feedstock and left for a longer period before being emptied. No mixing, stirring, and pumping required, low input in terms of process and mechanical demands, and low capital cost, but they are channeling and clogging and these types of reactors have larger volume and lower biogas yield. The methane production is commonly the highest at the beginning and decreases toward the end of the process as the substrate is being utilized [117, 140, 142].

5.6 Anaerobic baffled reactor (ABR)

Anaerobic baffled reactor (ABR) is a modification of Upper-flow Anaerobic Sludge-Bed Reactor. Anaerobic baffled reactor (ABR) initially gets the organic fraction of municipal solid waste (OFMSW) accompanied by decomposition process of the materials and eventually generates biogas by microorganisms' activities. This kind of reactor can possibly treat wastes with high solid content, and thus, it may be a viable alternative in some situations observed in developing countries. The raising contact time with the sludge (active biomass) results in treatment improvement. ABRs are powerful and able to treat a broad range of wastewater, but both remaining effluents and sludge still require additional treatment in order to be discharged or reused correctly [121, 143].

5.7 Hybrid bioreactor

Hybrid bioreactor represents the modern production of reactor with possibility to incorporate the benefits of both suspended solid and biofilm reactors. These types of reactors provide the benefits of the UASB concept related to the ones of the anaerobic filters, and nowadays can be considered more appropriate for the treatment of a sequence of soluble or partially soluble wastewater than other reactor systems. Hybrid reactor (combination of the basic types) and anaerobic baffled reactor (ABR) fall under this category [143–145].

Anaerobic digestion may consist of a single-stage operation, or a two-stage process. Single-stage operation is less efficient, but most commonly used because of its simplicity. Traditional single-stage digester is generally larger, and hence takes more energy to mix and heat compared to a two-stage digester; while, a two-stage digestion is more efficient overall compared to a single-stage process [50, 146–150]. Many different configurations and operational systems have been developed for anaerobic digesters for use in different applications. The goals normally are to shorten the start-up period, reduce operational instabilities, decrease washout of active biomass, and attempt to better accommodate the inevitable variations in feed composition. Operation, maintenance, and installation cost are other factors that substantially impact the economics of biogas generation.

Single-stage digesters are most typically utilized on account of their simplicity, but overall two-stage digesters are more effective. There is no specific digester kind can be recommended as being internationally appropriate. The selection in a given scenario has to consider a lot of factors involving the following: the prospects for disposal of the digestate and the effluent; nature and strength of the waste stream; the availability and skills level of the local workforce; local climatic conditions, infrastructural support and cost of energy; and the expense of construction and operation. Generation of biogas by AD is a helpful method to recover energy from organic waste, whereas considerably reducing the environmental effect of the waste [50]. In addition, the CSTR design is normally performed in single-stage systems, there the reactor operates, favoring both methanogenic and acidogenic microorganisms. These types of systems have lower capital and operating costs and are simple to operate, making them attractive for a broad range of applications through the last decades [151, 152]. Furthermore, the conversion of organic material to biogas is implemented during a series of biochemical reactions, which do not inevitably have the identical optimal environmental conditions. Single-stage digesters have simple design with less technical failure. In the other hand, it has higher retention time, and form foam and scum leading to potential failure [50, 117, 153]. In order to get higher reaction rates and hence a higher biogas yield, two- and multistage systems have developed to give optimal conditions for the various groups of microorganisms included in the degradation process [50, 153]. Four processes (hydrolysis, acidification, acetogenesis, and methanogenesis) in AD are separated in two-stage reactors. Thus, the first stage can be operated at lower pH, which is more favored for the growth of acidogenic and hydrolytic microorganisms; whereas, the second phase is operated to prefer the growth of methane forming microorganisms [154]. In the second step, the rate of limiting factor is normally the rate of microbial growth [155] since longer generation times for methane-producing archaea, and thus longer biomass retention times are required in this second stage, which in turn improves the biogas yield [118]. These kinds of digesters usually have a more steady performance than single-stage digesters, since they do not bear from the process disturbances caused by ammonia accumulation and the changes in the pH [155, 156]. Best phase extraction option can be given in multistage reactors, which can provide optimization and process control for each conversion point, leading to raised methane generation [50, 157]. Two-stage reactors increase in biomass digestion due to recirculation, it has constant feeding rate to methanogenic stage, and it is more robust and less susceptible to failure. In contrast, it has complex design and expensive to build and maintain, and solid particles need to be removed from the feedstock in the second stage [117].

6. Biogas technologies

There are undesired compounds and other gases contained in biogas are unwanted and are considered as biogas pollutants [11]. The concentrations of these impurities are dependent on the composition of the substrate from which the gas was produced [158]. The removal of these harmful components and other non-combustible gases makes biogas a more viable and economical alternative renewable energy source [96, 159]. The energy content of methane described by the Lower Calorific Value (LCV) is 50.4 MJ/kg CH_4 or 36 MJ/m^3 CH_4 (at STP conditions). Therefore, the higher the CO_2 or N_2 content is, the lower the LCV in biogas [11, 160]. Developing the quantity and quality of biogas often needs pretreatment to maximize methane yields and/or post-treatment to take out H_2S, which includes higher costs and considerable energy consumption. Therefore, scientific research has performed to develop a low-cost desulfurization process and improve AD conversion. Appealingly, there are a lot of techniques that have been approved to enhance the anaerobic digestion process, like pretreatment procedures using acidic/alkaline, ultrasonic, thermal methods [161–163]. Lately, there are various treatments targeting at get rid of the trace contaminants and undesired components from the biogas expanding its range of applications [11]. Biomethane involves two major treatment processes; cleaning and CH_4 enrichment (biogas upgrading). The cleaning of the biogas contains elimination of impurities and acidic gases; whereas, the enrichment process is for extraction of CO_2 from biogas [11, 96]. There are three major reasons for gas cleaning;

fulfill the requirements of gas appliances (gas engines, boilers, fuel cells, vehicles, etc.), increase the heating value of the gas, and standardization of the gas [58]. Biogas cleaning treatment process includes removal of undesired materials (such as, NH_3, siloxanes, H_2S, volatile organic compounds (VOCs), and CO) to increase the quality of biogas. However, it is practically only H_2S which is mainly targeted and many current biogas plants have H_2S elimination units normally rely on biological H_2S oxidation by aerobic sulfate oxidizing bacteria [11]. Biogas must be desulfurizated and also dried before usage to stop destroys the use of gas units. The concentration of H_2S between 100 and 3000 ppm in biogas generated by cofermentation of manure with harvesting debris or energy crops, in order to prevent an expensive deterioration of lubrication oil and excessive corrosion [21, 22]. CHPs are used for the utilization of biogas need generally levels of H_2S below 250 ppm. The existence of H_2S not only affects the quality and quantity of the biogas generated which can restrict its application, but also produces dangerous environmental emissions and corrodes the motors of biogas purification machinery [20, 23]. Nowadays, biological desulfurization process mainly used to remove of H_2S [21, 22]. Recent study conducted by Register Mrosso [164] reported that red rock (RR) is an available material for biogas purification which used to remove hydrogen sulfide from biogas [164]. The quality of raw biogas can be further improved via various upgrading techniques to remove the non-combustible components and as a result increasing the methane content to approximate natural gas quality (75–98% methane) [96]. Biogas has been upgraded to natural gas composition via methanation using renewable hydrogen [165]. The higher the methane content, the richer the biogas is in energy [12]. Biogas upgrading aims to increase the low calorific value of the biogas, and convert it to higher fuel standard [35]. In case the upgraded biogas is purified to specifications similar to natural gas, the final gas product is called biomethane [11, 166]. Biomethane is a gaseous fuel with physicochemical properties similar to those of natural gas, which makes it possible to inject it into the gas grid [96]. Currently, the specifications of the natural gas composition are depending on national regulations and in some countries >95% methane content is required [11].

Technological development plays an important role in biogas upgradation and purification processes in large-scale commercialization of biogas. There are various cleaning and upgrading techniques to improve the quality of raw biogas which can be categorized into physiochemical and biological technologies. Some of these techniques are conventional methods, including physical absorption, chemical absorption, membrane infiltration and biological methods, and others are considered as new technologies including cryogenic upgradation, membrane enrichment, multistage-, and high-pressurized AD [62, 96, 167, 168].

Physiochemical technologies for cleaning of biogas and its subsequent CH_4 enrichment can be grouped as follows: absorption process (physical and chemical absorption), Hybrid solution (mixed physical and chemical solvent), and physical separation (adsorption on solid surface; membrane; cryogenic) [96]. Novel technologies, such as cryogenic separation, in-situ upgrading, hydrate separation, and biological methods, represent the recent developments in biogas upgrading technologies. Biogas can be used as fuel for domestic stoves, boilers, internal engines, gas turbines, cars, and fuel cells, or injected into natural gas grids to replace gaseous fuel [35]. These techniques have been reported to yield biomethane typically containing 95–99% CH_4 and 1–3% CO_2. At this quality, the spectrum of applications for biogas widens, it can be used to serve the same applications as natural gas [96]. Gas upgrading and utilization as renewable vehicle fuel or injection into the natural gas grid is of increasing interest because the gas can be used in a more efficient way [21]. Types of upgrading plants are available in Sweden, and shows that around 70% of the biogas purification plants apply water-washing technologies [169].

6.1 Physiochemical technologies include

6.1.1 Physical absorption method

Physical absorption method uses water scrubbing system. Water scrubbing is the most commonly used technology for biogas cleaning and upgrading [170].

This process depends on the extraction of H_2S and CO_2 from the biogas because of their raised solubility in water compared to CH_4 (i.e., according to Henry's law, the solubility of CO_2 in water at 25°C is roughly 26 times higher compared to methane); whereas, physical absorption method is using organic solvents. This method relies on the same principle as water scrubbing; however, the absorption of CO_2 and H_2S is accomplished by the use of organic solvent instead of water.

6.1.2 Chemical absorption method

Various methods are used to bind the CO_2 molecules contained in the biogas, such as chemical scrubbers, utilize aqueous amine solutions (i.e., mono-, di-, or triethanolamine); chemical absorption method; and using amine solutions. One of the benefits of this technology is that H_2S can be totally absorbed in the amine scrubber. Amine scrubbing systems mostly contain a stripper and an absorber unit.

6.1.3 Pressure swing adsorption (PSA)

Pressure swing adsorption (PSA), which extracts the various gasses from biogas, relies on their molecular properties and the compatibility of the adsorbent matters. The adsorbents can be zeolites (Zeolite 13X, Zeolite 5A), carbon molecular sieve, activated carbon, and other substances with high surface area [171]. The major principle of PSA system depends on the properties of pressurized gasses to be appealed to solid surfaces. Thus, under high pressure, huge quantities of gas will be adsorbed, whereas, a decline of pressure will result in gas discharge. The PSA technology follows four different or equal duration stages, namely pressurization, adsorption, blow-down, and purge [171].

6.1.4 Membrane technology

Membrane technology is considered as an alternative to the traditional absorption-based biogas upgrading technology. The major principle of the membrane technology depends on the selective permeability characteristics of membranes allowing the biogas components to separate [172].

6.1.5 Cryogenic technique

The bases of this technology are the different liquefaction temperatures for biogas compounds [173]. It is conducted through a gradual decrease of biogas temperature allows the selective separation of CH_4 from both CO_2 and rest components. Thus, a high-purity biomethane is obtained in agreement with the quality standards for Liquefied Natural Gas (LNG). The easiest path to remove the impurities contained in biogas by means of cryogenic methods employs a constant pressure of 10 bar [9, 174–176]. The liquefaction is carried out by declining the temperature successively in order to get rid of each pollutant or mitigate them in different steps. The first step is often set up at −25°C, where mostly siloxanes, H_2O, and H_2S are obtained. A second set step is assigned at −55°C to partially liquefied CO_2, accompanied by a new decline until −85°C to totally get rid of the remaining CO_2 by a solidification step [177]. The liquefied CO_2 gained in the second temperature stage can be sold as high-purity by-products to improve the whole economic process performance. Another more normally used option contains a preparatory dry of the gas accompanied by a multistep compression up to 80 bar. This permits preserving a higher operational temperature of between −45 and −55°C, containing as major drawback a needful intermediate cooling in the multistep compression [178]. Cryogenic techniques represent a good option to be optimized because these techniques yield high-purity products, ranging between 95 and 99% [13, 179].

6.1.6 Chemical hydrogenation method

The reduction of CO_2 with H_2 can be either conducted biologically or chemically, based on Sabatier reaction. Regarding the chemical hydrogenation process, various catalysts, with Nickel and Ruthenium to be the most commonly used in industrial applications have already been tested under elevated temperature (e.g., 300°C) and pressure levels (e.g., 5–20 MPa) [180, 181]. Due to high selectivity, complete conversion of CO_2 and H_2 can be practically achieved [182]. Nevertheless, despite the high process efficiency, specific drawbacks still remain. For instance, the sustainability is affected by the presence of trace gasses in the biogas, which degenerate the catalysts leading to increased need for periodical replacement [183]. The high cost of energy to preserve the operational conditions, the lack of elements to synthesize effective catalysts, and the need for pure gasses are further technical challenges of the system [11].

6.2 Biological technologies

The biological biogas upgrading technologies are classified into chemoautotrophic and photosynthetic. Most of these configurations have been practically proven and are at an initial step of pilot or full scale application. The main benefit of such techniques is associated to the fact that the CO_2 is transformed into other energy containing or valuable added products at mild operational conditions (i.e., moderate temperature levels, atmospheric pressure) contributing extremely to a circular economy and sustainable bio-based.

6.2.1 Chemoautotrophic techniques

The chemoautotrophic biogas upgrading techniques rely on the action of hydrogenotrophic methanogens that can use H_2 to transform CO_2 to CH_4 depending on the following equation:

$$4H_2 + CO_2 \rightarrow CH_4 + H_2O \ \Delta G^\circ = -130.7 \ \text{kJ/mol} \tag{4}$$

But, in order to make the biological upgrading technology renewable, the necessary H_2 in the reaction has to be extracted from renewable source. Thus, the using renewable electricity concept for generation of H_2 by hydrolyzing water has attracted great attention, particularly in cases that residual electricity from solar panels or wind mills is exploited. Whereas, in the concept of in-situ biological biogas upgrading, H_2 is injected into a biogas digester in order to be connected with the endogenous CO_2, which is generated in the anaerobic reactor and be transformed into CH_4 by the action of autochthonous methanogenic archaea [11, 166].

6.2.2 Photoautotrophic methods

The photosynthetic biogas upgrading is an alternative technology to isolate the CO_2 in order to produce a CH_4-rich gas. By performing these techniques, H_2S elimination is further achieved; whereas, >54% of CO_2 is devoured. The methane recovery of photoautotrophic methods can reach up to roughly 97% relying on the reactor kinds and the selection of algal species.

Physicochemical methods are in general at high technology readiness levels, while biological methods are still new and not commercial yet. However, they offer huge potential in respect to feasibility, technological easiness, and potential. Biological upgrading opens new horizons for integrating different forms of renewable energy and besides upgrading can offer electricity storage advances and decoupling bioenergy production from biomass availability [11].

7. Biogas applications

Biogas generation serves three important functions: waste removal, environmental management, and energy production [12]. The first and most direct use of biogas is for heating and domestic purposes [184]. Biogas is an excellent fuel with a numerous application [62]. Biogas that is purified and enriched in methane can be used for household applications, automobile fuel (liquefied), or electricity generation [185, 186]. The biogas is mostly utilized as a combined heat and power (CHP) application in the overall world; and apart from it, it can be used in three sides such as fuels for cars, steam generation, and electric power. Biogas obtained from renewable organic waste is counted as an alternative energy for nonrenewable fuels due to its broad applications in fuel and transportation sector [62, 104, 187].

In general, Waste-to-Energy (WtE) technologies can be defined as any waste treatment processes that create energy from a waste source in any forms of energy carrier, i.e., electricity, heat, or transportation fuels [188]. Depending on a statement by World Energy Council, restricted landfilling capacities, rise in the quantity of produced waste, high costs of energy, and rising concerns of environmental issues are the summarized major factors for the growth in WtE market in the past decades. In 2013, the international WtE market encountered a growth of 5.5% and reached a value of 25.32 billion USD with respect to its previous year [50, 188]. Biogas is a flexible energy transporter, appropriate for various applications. One of the simplest applications of biogas is the immediate utilization for lighting, and cooking, but in a lot of countries biogas is currently utilized for combining heat and power generation (CHP) or it is upgraded and fed into natural gas grids, utilized in fuel cells or as car fuel [189]. Biogas is appropriate for production of electricity in combination with heat recovery. Normally, the gas is combusted in motors with internal combustion connected to turbine. The discharged heat (being about 60% of the used energy) is utilized for heating purposes for household requirements or maintenance of the anaerobic reactor. This method is broadly used for the treatment of activated sludge, debris generated from municipal wastewater treatment plants [184, 190, 191]. Electric power generation by gas turbines can be used by biogas as a fuel, hence substituting the natural gas for small-scale applications [184]. There is a large demand to make biogas transportable. This can be simply done only after taking out impurities such as CO_2, H_2S, and water vapor by compressing and filling the cylinders in it after scrubbing and drying processes [185]. Elimination of carbon dioxide from the flue gas assists to get fuel of higher calorific value as well as to remove the GHG [185, 192]. Biogas is an encouraging renewable source of energy. It can be immediately transformed into electricity, e.g., in a fuel cell, or burnt, discharging heat at high temperature, or burnt in a CHP for the simultaneous generation of heat and power, or fed into the natural gas network for energy rescuing purposes or it can used as fuel for cars, being sold by gas stations. Mostly, the biogas should be transported over long distances and must be purified before further utilization [18]. Biogas systems turn the cost of waste management into a revenue opportunity for farms, dairies, and industries. Converting waste into electricity, heat, or car fuel provides a renewable source of energy that can reduce dependence on foreign oil imports [189]. Biogas is mostly used in factory boilers and in engine generator sets to produce electricity and heat. In those cases, where an internal combustion engine is fuelled with biogas to produce electricity, the electricity can either be used by the facility itself or transferred to a local or national power grid [12]. The most profitable way to use biogas may be to convert it into natural gas. In reality, biogas can be utilized in all applications created for natural gas. The major difference between the two fuels is that, further to methane, natural gas consists of a variety of other hydrocarbons, like propane, butane, and ethane, which provide it a higher calorific value than pure methane. Biogas is normally burned in internal combustion motors to produce electric power. An electrical conversion efficiency of up to 25% can be obtained via small-scale internal combustion motors, with a rated capacity of less than 200 W as well as much higher electrical conversion efficiencies, of 30–35% can be provided through larger internal combustion motors (up to 1.5 MW). When biogas is utilized to generate electric power, there is the extra potential for heating water from the engine's exhaust and cooling systems. Combining hot water generation with electric power production can provide total conversion efficiency as high as 65–85%. An encouraging near-future application for electric power production is the utilization of gas turbines. Combined-cycle power stations are made up of waste heat recovery boilers, gas turbines, and steam turbines that function together to generate electric power in the larger-scale systems. Advanced gas turbine plants tend to be small, environment friendly, greatly efficient, and visually unobtrusive. Units as small as 200 kW are not uncommon, but only those greater than 800 kW have electrical conversion efficiencies that equal or surpass an internal combustion engine-based system. Gas turbines allow a greater fraction of waste heat to be recovered as steam, a critical commodity for many industries, so overall efficiency levels for gas turbines can be up to 75%. Recently, biogas applications are employed as fuel in fuel cells and as fuel for micro-CHP (combined heat and power). When connected with an organic Rankine cycle (ORC) turbine, a biogas-powered CHP can raise electrical efficiency by 8–10%, making total efficiency rate of 45–48% more than reasonable [12, 184]. Another very attractive application of biogas for electricity production is its use in fuel cells. The specialized cells for these purposes are described briefly by [193]. Identical efficiency rates are obviously being accomplished with biogas fuel cell technique. Sweden-based Acumentrics Corporation, for instance, has registered improved performances with its 5000 W fuel cells, known as solid oxide fuel cell (SOFC) systems, which work on biogas rather than hydrogen, which is difficult to handle, high cost, and difficult to store [12]. The utilization of biogas as a fuel for civil transport and road cars in place of natural gas is already widen in United States and Western Europe [194]. There are a lot of automobiles in Sweden turning on biogas in the urban public transport [184, 195]. Biogas is currently used in many developing countries as an alternative and renewable source of energy for wide spread range of applications. In contemporary times, biogas has been used most extensively in India and China. The Biogas Association in Germany, the world's largest producer country, included the three functions in its recent summary of what it called the national benefits of biogas production: 650 MW of installed electrical capacity comes from biogas, a reduction of 4 million tons per year of CO_2 emissions, revenues of $500 million for biogas farmers from electricity sales annually, and use by the AD process of biomass material that would otherwise end up in landfills. Economic production of biogas can be economically achieved for both large- and small-scale applications. Hence, it can be designed to fit into rural, urban, as well as regional and nationwide energy needs making it a versatile source of energy [12, 96, 107]. All over the world, Europe has registered the highest growth of biogas utilization with a notable 18% raise registered between 2006 and 2007. Sweden and Germany have registered the highest growth levels with Germany leading to brag over 4000 biogas plants, most of them are established on farms for electric power and heat co-generation [21, 96].

In Sweden, there is currently great interest in the biogas process, since it can stabilize and reduce various types of organic waste while producing renewable and environmentally friendly energy in the form of biogas. There is also increasing interest in both the production of biogas from municipal sewage treatment plants and on-farm biogas production within agriculture [196]. Biogas is also burned in boilers to produce hot water and steam in a variety of settings, including hotels, warehouses, factories, schools, prisons, and other public buildings. The forest-product segment is perhaps the largest user of biomass (combustion) energy in the industrial sector. In addition, in many countries, biogas is viewed as an environmentally attractive alternative to diesel and petrol for operating busses and other local transport vehicles [12]. The food and drinks industries are the largest users of AD for wastewater pretreatment. In 2006, 3400 GW of biogas power was generated in Germany, equivalent to 0.6% of the country's total energy consumption, reducing carbon dioxide emissions by 2.5 million tons. Countries such as Sweden is considered pioneered in the utilization of upgrading biogas as a light duty car fuel, and the use of biogas in the country has already exceeded natural gas [12, 197]. Identical attempts are also being made in Germany which presently turns on roughly 5000 anaerobic reactors for generation of bioenergy [198]. In the UK, fears over the utilization of biogas as fuel stem from an insufficiency of quality standards and infrastructure, as well as contest with other utilizes of biogas [199, 200]. At the moment, close to 50 biogas plants, mainly small, farm-scale ones, are in operation in Austria, Currently, around 25 biogas plants operate in Denmark, with capacities ranging from 50 to 500 tons of biomass feedstock per day. The resulting biogas is mainly used in heat and power generation applications, while the digested biomass is redistributed to farms as fertilizer. Swedish company Svensk Biogas has developed a passenger train that runs exclusively on biogas. The train has a range of 600 km and can attain speeds of 130 kph. There are also up to 100 municipal busses running on biogas. The Swedish agricultural sector is also increasingly using the residues from the anaerobic digestion of crops and clean organic waste in order to return nutrients to the soil and reduce its dependence on mineral fertilizers. Biogas can also be used to generate electricity alone or with heat (co-generation). Biogas can also be used, like pure methane, as a fuel for motor vehicles [12, 201]. Biogas can be considered as alternative green energy carrier for harnessing electricity, heat, and as a transport fuel [62, 202]. Biogas is a renewable source of energy that can be used as a substitute for natural gas or liquefied petroleum gas. Biogas is a clean, efficient, and renewable green source of energy, which can be used as a substitute for other fuels in order to save energy in rural areas [86, 203, 204]. A series of zero-waste technologies are presented. They are similar to the "five zeros" of the Olympic logo which are zero waste in the product life cycle, zero emissions, zero waste in activities, zero use of toxics, and zero resource waste. This design, firstly invented by Lakhal and H'Mida [205] was titled the Olympic Green Chain model. Lately, Khan and Islam [206] suggested a method for zero-waste (mass) utilization for an ideal urban setting, involving processing and regeneration of gas, liquid, and solid. In this process, kitchen sewage waste and waste are used for diverse purposes, involving generation of biogas, water heating from flue gas, good fertilizer for agricultural production and desalination. The carbon dioxide produced from biogas burning is used for the desalination plant. This process gets zero-waste in mass utilization. The technology development in this line has no negative impact on global warming. It is estimated that biogas usage in the world will be doubled in the coming years, increasing from 14.5 GW in 2012 to 29.5 GW in 2022 [37, 189, 207].

8. Advantages and limitations of biogas technologies

Biofuels are produced from biomass for a wide range of applications, such as cooking, heating, cooling, and transport. Biofuels can be solid (e.g., fuel-wood), liquid (e.g., bioethanol, biodiesel) or gaseous (e.g., biogas) [208, 209]. Biogas technology utilizes organic wastes for energy production, followed by recycling of the digested substrate as fertilizer [189]. Biogas can be used to generate heat or electricity, or as fuel for manufacturing or transport [210]. Electrical energy and heat generation from biogas is a source of green, environmentally friendly energy. At the same time, there is a reduction in methane emission from the decomposition of unmanaged biomass (especially animal droppings) [211–213]. Main benefits of biogas technology are to transform waste material into a valuable resource thus reducing waste, and providing valuable green energy [86, 189].

8.1 Advantages of biogas technologies

The production of biogas through anaerobic digestion (AD) offers significant advantages over other technologies of bioenergy production. It has been acknowledged as one of the most energy-efficient and green technology for bioenergy production [82]. For many reasons, it is a versatile renewable energy source [158], it can be produced when needed and can easily be stored [61], biogas can be easily upgraded to remove undesired components thus producing a higher fuel standard (Biomethane) with high specific caloric value [189]; it combines energy (gas) storage with generation [214]; the feedstock source is often a waste or problem product, and hence its use for energetic utilization resolves waste management problems [214]; biogas technology provides an excellent opportunity for mitigation of greenhouse gas emission, improving air quality, and reducing global warming [6, 215–217]. Biogas technology also has potential to mitigate climate change and eutrophication [218]; it can be used as an alternative to fossil fuels [158]; it can easily co-digest a range of feedstocks, thus providing an integrated waste management service; it provides valuable co-products such as nutrient-rich bioslurry [214]; biogas production is a treatment technology that generates renewable energy and recycles organic waste into a digested biomass, which can be used as fertilizer and soil amendment [82, 189]; Methane-rich biogas (biomethane) can also replace natural gas as a feedstock for producing chemical materials [82]; biogas is considered to be the future of renewable and sustainable energy [219]; noise levels generated by methane-powered engines are considerably lower than those of diesel engines, a plus in congested urban environments [12]; biogas technology has an important role to play in the waste management, renewable energy, water, and nutrient (food security) sectors [214]. The development of a national biogas sector contributes to increase the income in rural areas and creates new jobs [189].

The benefits of using co-digestion techniques for optimizing biogas production yields which including dilution of potential toxic compounds, improved balance nutrients, synergistic effect of microorganism, increased load of biodegradable organic matter, and higher biogas yield [82, 220, 221]. In small-scale installations, worldwide, the gas is primarily utilized for lighting and cooking. In larger units, the gas can be used for co-generation (generation of heat and electricity), as vehicle fuel or as fuel in industrial processes [58].

8.2 Disadvantages and problems

Biogas production from anaerobic digestion (AD) suffers from several technical limitations. The social acceptance of biogas is usually hampered by health and environmental concerns. There are undesired and harmful substances contained in biogas which considered as biogas pollutants (such as H_2S, Si, volatile organic compounds (VOCs), siloxanes, CO, and NH_3). H_2S and NH_3 are toxic and extremely corrosive, damaging the combined heat and power (CHP) unit and metal parts via emission of SO_2 from combustion [11, 160, 215]. The existence of H_2S not only influences the quantity and quality of the biogas produced which can limit its application, but also generates harmful environmental emissions and corrodes the engines of biogas purification machinery [23, 163]. It also causes excessive corrosion and expensive deterioration of lubrication oil [21]. Moreover, the presence of siloxanes in biogas, even in minor concentrations, is associated with problems. It is well known that during combustion silicone oxides generate sticky residues, which deposit in biogas combustion engines and valves causing malfunction [11, 160]. Biogas produced by AD still contains impurities. Therefore, the systems used in the production of biogas are not efficient [189]. The quality and quantity of biogas usually requires pretreatment to maximize methane yields and post-treatment to remove H_2S, which involves considerable energy consumption and higher costs [163]. There are no new technologies yet to simplify the process and make it abundant and low cost. Similar to other renewable energy sources (e.g., solar, wind) production of biogas is also influenced by the climate. The optimal temperature required for bacteria to digest waste is about 37°C. In cold weather, digesters need heat energy to preserve a fixed biogas supply [189]. The greatest challenge encountering the utilization of biogas as a truck and bus fuel has been the restricted driving range that it provides, meaning that drivers must refuel much more often than they would in petrol- or diesel-powered cars [12].

9. Laws and guidelines concerning biogas plants

Most countries are in the process of instituting legislation to regulate the biogas industry [12]. Mostly, all parts of the plant must be checked out and licensed by the authorities. This involves installations such as tanks for liquid manure, bioreactors, gasholders, ignition oil tanks, stores, and combined heat and power stations (CHP). In biogas plants, the formation of explosive gas mixtures can happen. Thus, a system for plant security has to be present relating to installation and operation of electrical instruments in regions with high risk of explosions [18].

In Africa and the Middle East, the alternative energy market is new, so there are few government regulations and formal incentives. With inexpensive and abundant energy sources—coal in South Africa and oil in the Middle East—interest on the part of the state sector in renewable energy has been minimal. Although there are few laws or regulations pertaining to the biogas sector—reflecting the undeveloped state of the industry, due primarily to abundant and inexpensive sources of traditional fossil fuels, which gives little or no incentive to look for alternative energy sources, South Africa is leading the way in the region.

The South African Minerals and Energy Department, for instance, published its energy efficiency and renewable strategy statement in 2006, which involved statements and targets on biogas. These involved a "target requirement" to generate 4% nearly 10,000 GWh of the country's electric power from renewable sources in 2013. In some of African countries, involving Egypt, South Africa, and Morocco, there are disagreement of interest over who has the lawful right to utilize municipal and common, specially maize (corn), or tribal, land to cultivate biogas crops [12].

In Europe, legislation is well developed, reflecting the relatively high level of biogas production in many EU markets, such as Germany and the UK, where biogas is the fastest-growing segment of the renewable fuels industry [12]. Europe is the most advanced biogas market in the world and legislation is much more developed than in other regions, with laws and regulations that not only relate to requirements to treat organic waste in certain ways, but also reach an unusual level of detail regarding machine, plant, and process specifications [18]. The European Commission is proposing an increase in the use worldwide of renewable energy sources to 20% of the total demand, which would limit global temperature changes to no more than 2°C above pre-industrial levels. The Commission has stated, rather vaguely, that it seeks, via both voluntary and legal means, to improve the EU's energy efficiency by 20%, in order to "make the EU the most energy-efficient region in the world" [12].

In Germany, there are a lot of state laws, regulations, norms, and guidelines of branch institutions were released in order to preserve a secure and smooth biogas plants operation. These cover their waste management, installation, operation, and supply. A lot of offices are included in managerial decision for the a plant construction, e.g., the office for noise control-traffic-energy-climatic protection, the planning department and building control office, the water regulatory authority, the natural conservation authority, food control, the authority for nutrition, the authority of agriculture, the office for veterinary matters, the office for technology and plant safety, the public order office, etc. [18]. In Germany, for instance, biogas is the fastest-growing segment in the alternative energy industry. Thus, both local and national governments are beginning to oversee the sector at a much more detailed level than before. Laws requiring more cooperation between biogas producers and public utility companies, to avoid electrocution of workers who may shut down power to an electric cable but the farm-based generator continues to feed energy into it, have come into force in most states since 2006. Most states now have laws requiring electricity utilities to buy excess biogas production, either via established gas distribution networks or directly through national pipelines. EU energy vision includes a cut in carbon dioxide emissions by at least 20% by 2020 [12]. The German law of biowaste biomass specifies the biomasses and the technical processes supported. Approved biomass consists of pure herbaceous products. Not approved are fossil fuels, mixed wastes, mud, sewage sludge, and port sludge.

Debris from biogas plants are undergo to the German law of fertilizers, involving when domestic waste water undergo fermentation or when it is blended with agricultural substrates. The German law of fertilizers control single guidelines like temporary permissions to use fertilizer, the determination of fertilizing requirements, the maximum limits for the utilized amount of fertilizer, methods to use, and much more. The given name of these regulations is the "principles of the good and professional execution of fertilizing." In Germany, no liquid secondary raw material fertilizer or liquid manure are permitted to be utilized from November 15th to January 15th. This is due to the soil is chilled and the product of fermentation cannot go through the soil. The distribution of debris from biogas plants is also managed. Maximum amount of total nitrogen from industrial fertilizer are allowed on pasture land is not more than 210 kg/ha.a and on ground used agriculturally is not more than 170 kg/ha.a. For phosphates, 120 kg/ha.a as maximum limits for both on pasture as well as on ground used agriculturally, for calcium the limit is regulated to 360 kg/ha.a. Farms with greater than 10 ha of ground utilized agriculturally are compelled to make a written fertilizer balance to keep track of their utilization [18].

In biogas plants, substances able to contaminate water are treated. In most countries, it is banned by law to contaminate water. Best available measures have to be taken in biogas plants to preserve water from pollution. The distribution of debris in agriculture should be done as stated in specified techniques [18].

In North America, the main US biogas legislation is the Biogas Production Incentives Act of 2007. A widespread unwillingness on the part of electricity companies in both the US and Canada to cooperate with biogas farm-scale producers has discouraged the development of the biogas sector. In response, many states are introducing legislation to oblige electricity providers to work with biogas producers and to buy any excess electricity. Electricity utilities, though, are generally unhappy with the arrangement, as the electricity produced by the biogas plant to run the farm is considered by them to be lost revenue. In Canada, the electric power prices provided to farmers generating biogas are still not enough to make production feasible. The Standard Offer Contract (SOC) program lately launched in Ontario, though, whereas not providing immediate financial inducements, is a first stage in the direction of promoting biogas generation though the utilization of energy crops and anaerobic reactors [12].

In Latin America, laws on quality standards for agricultural and organic waste are now being introduced, although in most cases these have not yet been clearly defined. Governments are only now beginning to make use of tax and investment incentives to encourage production. Latin America Argentina, Brazil and Colombia have the most developed legislative frameworks for the biofuels industry overall, each having set minimum requirement levels for the percentage of renewable fuels in petrol, diesel oil, and fuel oil. In Argentina, for example, the Biofuels Act of May 2006 was the first law to grant tax incentives to alternative fuel producers. The law will initially be in effect for 15 years. It stipulates that biogas can be produced from raw materials in the agricultural, agro-industrial or organic waste sectors provided they meet the government's quality standards which have yet to be clearly identified [12].

In Asia, Asian governments are beginning to encourage the biogas sector, basically by providing financial and legal incentives to produce energy from organic waste sources in large municipal waste dumps, the existence of which is coming under heavy local opposition in many cities. In India and Mongolia, household and industrial waste laws were legislated in early 2007, requiring the separation and treatment of both kinds of waste at an early processing stage, with a view to using organic and biodegradable waste for energy production. Recently, the Indonesian government issued a statute requiring minimum levels of biogas production from new waste disposal sites that are under construction. The way in which the gas is produced, whether it be by composting, landfill, or anaerobic digestion, can be selected by the producer [12].

10. Conclusions

Most of industrial and chemical processes produce wastes. Raw materials for producing biogas by anaerobic digestion are biomass feedstocks which include; municipal solid waste (MSW), industrial solid wastes and industrial wastewaters, food waste, livestock manure, sewage sludge, agricultural manures, catch crops, energy crops, and microalgae. Biogas production serves three important functions: waste removal, environmental management, and energy production. Biogas is a versatile renewable green energy source, which can be used for replacement of fossil fuels in power and heat production, and as gaseous vehicle fuel. Biogas technology is considered an alternative green energy resource. The use of energy and manure can lead to social economic benefits, green environment, and also contributes toward sustainable development. Renewable energy technologies provide an excellent opportunity for mitigation of greenhouse gas emission and reducing global warming through substituting conventional energy sources. Anaerobic digestion of organic waste is the most desirable management method, and this research discussed it in detail. There are undesired compounds and other gases contained in biogas which are considered as biogas pollutants. Improving the quality and quantity of biogas usually requires pretreatment to maximize methane yields and/or post-treatment to remove H_2S. The pretreatment of organic waste is the key process step in biogas production plants. Biomethane involves two major treatment processes; cleaning and CH_4 enrichment (biogas upgrading). The cleaning of the biogas consists of removal of acidic gases and impurities, while the enrichment process is for separation of CO_2 from biogas. It is noted that, there are different kinds of digesters, typically digester type is to be selected depending on the characteristics of the major feedstock used, particularly total solid. The study also concluded that, there are various cleaning and upgrading techniques to improve the quality of raw biogas which can be categorized into physiochemical and biological technologies. Some of these techniques are conventional methods, including physical absorption, chemical absorption, membrane infiltration, and biological methods, and others are considered new technologies, including cryogenic upgradation, membrane enrichment, multistage-, and high-pressurized AD. Novel technologies, such as cryogenic separation, in-situ upgrading, hydrate separation, and biological methods, represent the recent developments in biogas upgrading technologies. The biological biogas upgrading technologies can be classified into chemoautotrophic and photosynthetic. Physicochemical methods are in general at high technology readiness levels, while biological methods are still new and not commercial yet. It is reported that, most countries are in the process of instituting legislation to regulate the biogas industry. Europe is the most advanced biogas market in the world and legislation is much more developed than in other regions. Biogas is considered to be the future of renewable and sustainable energy.

Author details

Issa Korbag[1*], Salma Mohamed Saleh Omer[1], Hanan Boghazala[2]
and Mousay Ahmeedah Aboubakr Abusasiyah[3]

1 Department of Chemistry, Faculty of Arts and Sciences, University of Benghazi, Kufra, Libya

2 Department of Botany, Faculty of Arts and Sciences, University of Benghazi, Kufra, Libya

3 Department of Botany and Microbiology, Faculty of Science, Sebha University, Libya

*Address all correspondence to: issa.omar@uob.edu.ly

References

[1] Chaiprasert P. Biogas production from agricultural wastes in Thailand. Journal of Sustainable Energy & Environment. 2011. p. 63-65. Available from: http://jseejournal.com/journal.php?id=11

[2] Rai M, Da Silva SS. Nanotechnology for Bioenergy and Biofuel Production. Green Chemistry and Sustainable Technology. Cham, Switzerland: Springer; 2017. Available from: https://doi.org/10.1007/978-3-319-45459-7

[3] Maciejewska A, Veringa H, Sanders J, Peteves SD. Co-Firing of Biomass with Coal: Constraints and Role of Biomass Pre-Treatment, DG JRC Institute for Energy Report, EUR 22461 EN; 2006

[4] Van Loo S, Koppejan J. The handbook of biomass combustion and co-firing. London: Earth Scan; 2008

[5] Lee S, Shah YT. Biofuels and Bioenergy: Process and Technologies. Boca Raton FL: CRC Press; 2013. pp. 10-20

[6] Panwar NL, Kaushik SC, Kothari S. Role of renewable energy sources in environmental protection: A review. Renewable and Sustainable Energy Reviews. 2011;**15**(3):1513-1524

[7] Demirbas A. Methane Gas Hydrate: as a Natural Gas Source. In: Methane Gas Hydrate. Green Energy and Technology. 1st ed. London: Springer; 2010

[8] Ngan NVC et al. Anaerobic digestion of rice straw for biogas production. In: Gummert M, Hung N, Chivenge P, Douthwaite B, editors. Sustainable Rice Straw Management. Cham: Springer International Publishing; 2020. pp. 65-92

[9] Muñoz R, Meier L, Diaz I, Jeison D. A review on the state-of-the-art of physical/chemical and biological technologies for biogas upgrading. Reviews in Environmental Science and Biotechnology. 2015;**14**:727-759

[10] Petersson A, Wellinger A. Biogas upgrading technologies developments and innovations. Technical report, Task 37. Ireland: IEA Bioenergy; 2009. p. 20

[11] Angelidaki I, Treu L, Tsapekos P, Luo G, Campanaro S, Wenzel H, et al. Biogas upgrading and utilization: Current status and perspectives. Biotechnology Advances. 2018;**36**(2):452-466. DOI: 10.1016/j.biotechadv.2018.01.011

[12] Ford S. Advances in Biogas. Leatherhead, UK: Pira International Ltd; 2007. pp. 13-14

[13] Baena-Moreno FM, Rodríguez-Galán M, Vega F, et al. Biogas upgrading by cryogenic techniques. Environmental Chemistry Letters. 2019a;**17**:1251-1261. Available from: https://doi.org/10.1007/s10311-019-00872-2

[14] Montingelli ME, Tedesco S, Olabi AG. Biogas production from algal biomass: A review. Renewable and Sustainable Energy Reviews. 2015;**43**:961-972. DOI: 10.1016/j.rser.2014.11.052

[15] Aguirre-Villegas HA, Larson R, Reinemann DJ. Effects of management and co-digestion on life cycle emissions and energy from anaerobic digestion. Greenhouse Gases: Science and Technology. 2015;**5**:603-621. DOI: 10.1002/ghg.1506

[16] Chatterjee P, Ghangrekar MM, Rao S. Low efficiency of sewage treatment plants due to unskilled operations in India. Environmental Chemistry Letters. 2016;**14**:407-416. DOI: 10.1007/s10311-016-0551-9

[17] Sahota S, Shah G, Ghosh P, et al. Review of trends in biogas upgradation technologies and future perspectives. Bioresource Technology Reports. 2018;**1**:79-88. DOI: 10.1016/j.biteb.2018.01.002

[18] Deublein D, Steinhauser A. Biogas from Waste and Renewable Resources. 1st ed. Weinheim: Wiley-VCH; 2008

[19] Huber S, Mair K. Untersuchung zur Biogas zusammensetzung bei Anlagen aus der Landwirtschaft. Augsburg: Ergebnisbericht des Bayerischen Landesamts f ü r Umweltschutz; 1997

[20] Khan IU, Othman MHD, Hashim H, Matsuura T, Ismail AF, Arzhandi MRD, et al. Biogas as a renewable energy fuel – A review of biogas upgrading, utilization and storage. Energy Conversion and Management. 2017;**150**:277-294

[21] Weiland P. Biogas production: Current state and perspectives. Applied Microbiology and Biotechnology. 2010;**85**:849-860

[22] Schneider R, Quicker P, Anzer T, Prechtl S, Faulstich M. Grundlegende Untersuchungen zur effektiven, kostengünstigen Entfernung von Schwefelwasserstoff aus. In: Biogas Biogasanlagen Anforderungen zur Luftreinhaltung. Augsburg: Bayerisches Landesamt für Umweltschutz; 2002

[23] Lar JS, Xiujin LI. Removal of H_2S during anaerobic bioconversion of dairy manure. Chinese Journal of Chemical Engineering. 2009;**17**:273-277. DOI: 10.1016/S1004-9541(08)60205-0

[24] Sakhawat A, Naseem Z, Zahida N, Shumaila U. Impact of biogas technology in the development of rural population. Pakistan Journal of Analytical & Environmental Chemistry. 2013;**14**(2):65-74

[25] World Energy Council. New Renewable Energy Resources: A Guide to the Future. London: Kogan Page Limited; 1994

[26] Buren V. A Chinese Biogas Manual. London: Intermediate Technology Publications Ltd.; 1979. p. 11

[27] Ministry of New and Renewable Energy. Booklets on Renewable Energy. India: MNRE; 2011. Available from: http://mnre.gov.in/re-booklets.htm [Accessed: 04 May 2011]

[28] Wang LK, Hung Y-T, Lo HH, Yapijakis C, Li KH, editors. Handbook of Industrial and Hazardous Wastes Treatment. 2nd ed. New York-Basel: Marcel Dekker, Inc.; 2004

[29] Mshandete AM, Parawira W. Biogas technology research in selected sub-Saharan African countries - A review. African Journal of Biotechnology. 2009;**8**(2):116-125

[30] FAO. A system approach to biogas technology. In: Biogas Technology: A Training Manual for Extension. Nepal: FAO/CMS; 1996

[31] Azhar N, Anwar B. Biogas Production from Vegetable Waste at Thermophilic Conditions. Lahore: PEC; 2012. p. 67

[32] Seadi T, Rutz D, Prassl H, Kottner M, Finsterwalder T, Volk S, et al. Biogas Handbook. Esbjerg, Denmark: University of Southern Denmark; 2008. p. 11

[33] Lau C, Tsolakis A, Wyszynski M. Biogas upgrade to syn-gas (H_2–CO) via dry and oxidative reforming. International Journal of Hydrogen Energy. 2011;**36**(1):397-404

[34] Tippayawong N, Thanompongchart P. Biogas quality upgrade by simultaneous removal of CO_2 and H_2S in a packed column reactor. Energy. 2010;**35**(12):4531-4535

[35] Sun Q, Li H, Yan J, Liu L, Yu Z, Yu X. Selection of appropriate biogas upgrading technology - A review of biogas cleaning, upgrading and utilisation. Renewable and Sustainable Energy Reviews. 2015;**51**:521-532

[36] Maroneze MM, Zepka LQ, Vieira JG, Queiroz MI, Jacob-Lopes E. Production and use of biogas in Europe: A survey of current status and perspectives. Revista Ambiente E Água. 2014;**9**(3):445-458. DOI: 10.4136/1980-993X

[37] Kárászová M, Sedláková Z, Izák P. Gas permeation processes in biogas upgrading: A short review. Chemical Papers. 2015;**69**(10):1277-1283

[38] Balcioglu H, EL-Shimy. M., and Soyer. K. Renewable Energy – Background. Economics of Variable Renewable Sources for Electric Power Production. Germany: Lambert Academic Publishing/Omniscriptum Gmbh & Company Kg; 2017

[39] Owusu PA, Asumadu-Sarkodie S. A review of renewable energy sources, sustainability issues and climate change mitigation. Cogent Engineering. 2016;**3**(1):1167990. DOI: 10.1080/23311916.2016.1167990

[40] Demirbas A. Recent advances in biomass conversion technologies. Energy Education Science and Technology. 2000;**6**:19-40

[41] Sharma A, Srivastava J, Kumar A. A comprehensive overview of renewable energy status in India. In: Thangavel P, Sridevi G, editors. Environmental Sustainability. New Delhi: Springer; 2015

[42] Rathore NS, Panwar NL. Renewable Energy Sources for Sustainable Development. New Delhi, India: New India Publishing Agency; 2007

[43] Asumadu-Sarkodie S, Owusu PA. A review of Ghana's energy sector national energy statistics and policy framework. Cogent Engineering. 2016b;**3**:1-27. DOI: 10.1080/23311916.2016.1155274

[44] Asumadu-Sarkodie S, Owusu PA. Carbon dioxide emissions, GDP, energy use and population growth: A multivariate and causality analysis for Ghana, 1971-2013. Environmental Science and Pollution Research International. 2016c;**23**:13508-13520. DOI: 10.1007/s11356-016-6511-x

[45] International Energy Agency (IEA). Renewables Information 2013 - IEA Statistics. Paris: IEA; 2013a

[46] Akan MÖ, Selam AA, Fırat SÜ. Renewable energy sources: Comparison of their use and respective policies on a global scale. In: Erdoğdu M, Arun T, Ahmad I, editors. Handbook of Research on Green Economic Development Initiatives and Strategies. Hershey, PA: IGI Global; 2016. pp. 238-269. DOI: 10.4018/978-1-5225-0440-5.ch011

[47] Prasad S, Sheetal KR, Venkatramanan V, Kumar S, Kannojia S. Sustainable energy: Challenges and perspectives. In: Shah S et al., editors. Sustainable Green Technologies for Environmental Management. Singapore: Springer; 2019a. pp. 175-197. DOI: 10.1007/978-981-13-2772-8_9

[48] Farrell AE, Gopal AR. Bioenergy research needs for heat, electricity, and liquid fuels. MRS Bulletin. 2008;**33**:373-380

[49] Beckett M. Biotechnology's Role in Sustainable Energy for All. 2012. Available from: http://www.pub.ac.za/files/Biotech%20and%20Sustainable%20energy.pdf

[50] Meisam Tabatabaei HG. Biogas: Fundamentals, Process and Operation. Biofuel and Biorefinery Technologies. Vol. 6. Cham, Switzerland: Springer; 2018. pp. 1-230

[51] Habib K, Schmidt JH, Christensen P. A historical perspective of global warming potential from municipal solid waste management. Waste Management. 2013;**33**(9):1926-1933. DOI: 10.1016/j.wasman.2013.04.016 (Elsevier Ltd)

[52] Flotats X. Biogas: Perspectives of an old technology. In: Bastidas-Oyanedel JR, Schmidt J, editors. Biorefinery. Cham: Springer; 2019

[53] Bux F, Chisti Y. Algae Biotechnology: Products and Processes. Green Energy and Technology. Cham: Springer International Publishing; 2016

[54] Zabed HM, Akter S, Yun J, Zhang G, Zhang Y, Qi X. Biogas from microalgae: Technologies, challenges and opportunities. Renewable and Sustainable Energy Reviews. 2020;**117**:109503. DOI: 10.1016/j.rser.2019.109503

[55] Ward AJ, Lewis DM, Green FB. Anaerobic digestion of algae biomass: A review. Algal Research. 2014;**5**:204-214

[56] Williams PJLB. Biofuel: Microalgae cut the social and ecological costs. Nature. 2007;**450**:478

[57] Zabed HM, Akter S, Yun J, Zhang G, Awad FN, Qi X, et al. Recent advances in biological pretreatment of microalgae and lignocellulosic biomass for biofuel production. Renewable and Sustainable Energy Reviews. 2019;**105**:105-128

[58] Persson M, Jönsson O, Wellinger A. Biogas upgrading to vehicle fuel standards and grid injection. Technical Report, Task37: Energy from biogas and Lanfill Gas. Paris, France: IEA Bioenergy; 2006. Available from: https://www.ieabioenergy.com/publications/biogas-upgrading-to-vehicle-fuel-standards-and-grid-injection/

[59] Menzi H. Manure management in Europe: Results of a recent survey. In: Proceedings of the 10th Conference of the FAO/ESCORENA Network on Recycling Agricultural, Municipal and Industrial Residues in Agriculture (RAMIRAN); 14-18 May; Strbske Pleso, Slovak Republic. 2002. pp. 93-102

[60] FAOSTAT – Food and Agriculture Organization of the United Nations. Rome, Italy: FAO Statistical Databases; 2003. Available from: http://faostat.fao.org/

[61] Nielsen JBH, Seadi TA, Popiel PO. The future of anaerobic digestion and biogas utilization. Bioresource Technology. 2009;**100**(22):5478-5484

[62] Chozhavendhan S, Gnanavel G, Karthiga DG, Subbaiya R, Praveen Kumar R, Bharathiraja B. Enhancement of feedstock composition and fuel properties for biogas production. In: Praveen Kumar R, Bharathiraja B, Kataki R, Moholkar V, editors. Biomass Valorization to Bioenergy. Energy, Environment, and Sustainability. Singapore: Springer; 2020. pp. 113-131

[63] Fujino J, Morita A, Matsuoka Y, Sawayama S. Vision for utilization of livestock residue as bioenergy resource in Japan. Biomass and Bioenergy. 2005;**29**:367-374

[64] Muller JA, Winter A, Strunkmann G. Investigation and assessment of sludge pre-treatment processes. Water Science and Technology. 2004;**49**:97-104

[65] Schnürer A, Jarvis A. Microbiological Handbook for Biogas Plants. Swedish Waste Management, Reports from Avfall Sverige; U2009:03; 2010

[66] Treichel H, Fongaro G. Improving biogas production: Technological challenges, alternative sources, future developments. In: Biofuel and Biorefinery Technologies (Book 9). 1st ed. Cham, Switzerland: Springer; 2019

[67] Barros RM, Filho GLT, Santos AHM, Ferreira CH, Pieroni MF, Moura JS, et al. A potential of the biogas generating and energy recovering from municipal solid waste. Renewable Energy Focus. 2018;**25**:4-16

[68] Rahman AA. Biogas energy - An alternative solution for sustainable energy in rural areas of Bangladesh [Master's thesis]. Sweden: Department of Environmental Science, Lund University; 2000

[69] Agustini CA, Spier F, Costa M, Gutterres M. Biogas production for anaerobic co-digestion of tannery solid wastes under presence and absence of the tanning agent. Resources, Conservation and Recycling. 2018;**130**:51-59

[70] Priebe GPS, Kipper E, Gusmão AL, Marcilio NR, Gutterres M. Anaerobic digestion of chrome-tanned leather waste for biogas production. Journal of Cleaner Production. 2016;**129**:410-416

[71] Mannucci A, Munz G, Lubello C. Anaerobic treatment of vegetable tannery wastewaters: A review. Desalination. 2010;**264**:1-8

[72] Dyah S, Sriharti. Biogas development: Dissemination and barriers. In: IOP Conference Series: Earth and Environmental Science. Vol. 277. 2019. p. 012018

[73] Abubakar BSU, Ismail N. Anaerobic digestion of cow dung for biogas production. ARPN Journal of Engineering and Applied Sciences. 2012;7(2):169-172

[74] Abbasi T, Abbasi SA, Tauseef SM. Biogas Energy. New York/Dordrecht/Heidelberg/London: Springer Science; 2012

[75] Mudhoo A. Anaerobic digestion: Pretreatments of substrates. In: Mudhoo A, editor. Biogas Production: Pretreatment Methods in Anaerobic Digestion. New Jersey: Wiley; 2012. p. 199e212

[76] Venkata Mohan S, Lalit Babu V, Sarma PN. Effect of various pretreatment methods on anaerobic mixed microflora to enhance biohydrogen production utilizing dairy wastewater as substrate. Bioresource Technology. 2008;**99**:59-67

[77] Weiland P. Trockenfermentation in der Landwirtschaft-Welche Substrate und Techniken finden Anwendung. In: Bilitewski B, Werner P, Dornack C, Stegmann R, Rettenberger G, Faulstich M, Wittmaier M, editors. Anaerobe biologische Abfallbehandlung. Dresden: Forum für Abfallwirtschaft und Altlasten; 2008a. pp. 235-245

[78] Braun R. Anaerobic digestion: A multi-faceted process for energy, environmental management and rural development. In: Ranalli P, editor. Improvement of Crop Plants for Industrial End Uses. Dordrecht: Springer; 2007. pp. 335-415

[79] Baserga U. Landwirtschaftliche Co-Vergärungs-biogasanlagen, Biogasaus organischen Reststoffen und Energiegras, FAT-Berichte No. 512. Tänikon, Switzerland; Eidgenössische Forschungsanstalt für Agrarwirtschaft und Landtechnik (FAT); 1998

[80] Agler MT et al. Waste to bioproduct conversion with undefined mixed cultures: The carboxylate platform. Trends in Biotechnology. 2011;**29**(2):70-78

[81] Rapport J, Zhang R, Jenkins BM, Williams RB. Current Anaerobic Digestion Technologies Used for Treatment of Municipal Organic Solid Waste. Contractor Report to the California Integrated Waste Management Board. Department of Biological and Agricultural Engineering, University of California: Davis, CA; 2008

[82] Turco M, Ausiello A, Micoli L. Treatment of Biogas for Feeding High Temperature Fuel Cells: Removal of Harmful Compounds by Adsorption Processes. Cham, Switzerland: Springer; 2016

[83] Fehrenbach H, Giegrich J, Reinhardt G, Sayer U, Gretz M, Lanje K, et al. Kriterien einer nachhaltigen Bioenergienutzung im globalen Maßstab. Heidelberg: Institute fürEnergieund Umweltforschung, UBA-Forschungsbericht. 2008;**206**:41-112

[84] Adarme OFH, BEL B, DRS L, LVA G, de Aquino SF. Methane and hydrogen production from anaerobic digestion of soluble fraction obtained by sugarcane bagasse ozonation. Industrial Crops and Products. 2017;**109**:288-299

[85] Saratale RG, Kumar G, Banu R, Xia A, Periyasamy S, Saratale GD. A critical review on anaerobic digestion of microalgae and macroalgae and co-digestion of biomass for enhanced methane generation. Bioresource Technology. 2018;**262**:319-332

[86] Rajendran K, Aslanzade S, Taherzadeh J. Household biogas digesters – A review. Energies. 2012;**5**:2911-2942. DOI: 10.3390/en5082911

[87] NAS. Methane Generation from Human, Animal and Agricultural Waste. Washington, DC, USA: National Academy of Sciences; 1977

[88] Meyer A, Ehimen E, Holm-Nielsen J. Future European biogas: Animal manure, straw and grass potentials for a sustainable European biogas production. Biomass Bioenergy. 2018;**111**:154-164

[89] Lee S, Speight JG, Loyalka SK. Handbook of Alternative Fuel Technology. Boca Raton, FL: CRC Press; 2007

[90] Thornley P, Adams P. Biomass conversion technologies. In: Greenhouse Gas Balances of Bioenergy Systems. USA: Academic Press; 2017. pp. 107-139

[91] Weiland P. Production and energetic use of biogas from energy crops and wastes in Germany. Applied Biochemistry and Biotechnology. 2003;**109**:1-2

[92] Pagés-Díaz J et al. Co-digestion of different waste mixtures from agro-industrial activities: Kinetic evaluation and synergetic effects. Bioresource Technology. 2011;**102**(23):10834-10840

[93] Pagés-Díaz J et al. Semi-continuous co-digestion of solid cattle slaughterhouse wastes with other waste streams: Interactions within the mixtures and methanogenic community structure. Chemical Engineering Journal. 2015;**273**:28-36

[94] Meinhold J, Pedersen H, Arnold E, Issacs S, Henze M. Effect of continuous addition of an organic substrate to the anoxic phase on biological phosphate removal. Water Science and Technology. 1998;**38**:97-105

[95] Bonmati A, Flotats X. Air stripping of ammonia from pig slurry: Characterization and feasibility as a pre- or post-treatment to mesophilic anaerobic digestion. Waste Management. 2003;**23**:261-272

[96] Kigozi R, Muzenda E, Aboyade AO. Biogas technology: Current trends, opportunities and challenges. In: Proceedings of 6th International Conference on Green Technology, Renewable Energy and Environmental Engineering (ICGTREEE'2014); 24-28 November 2014; Cape Town, South Africa. pp. 311-317

[97] Kadam R, Panwar NL. Recent advancement in biogas enrichment and its applications. Renewable and Sustainable Energy Reviews. 2017;**73**:892-903

[98] Chen C, Zheng D, Liu GJ, Deng LW, Long Y, Fan ZH. Continuous dry fermentation of swine manure for biogas production. Waste Management. 2015;**38**:436-442

[99] Van Stephan F, Mathot M, Decruyenaere V, Loriers A, Delcour A, Planchon V, et al. Consequential environmental life cycle assessment of a farm-scale biogas plant. Journal of Environmental Management. 2016;**175**:20-32

[100] Viessman W Jr, Hammer MJ. Water Supply and Pollution Control. New York: Harper Collins College Publishers; 1993. pp. 513-679

[101] Steadman P. Energy, Environment and Building: A Report to the Academy of Natural Sciences of Philadelphia. Cambridge, London: Cambridge University Press; 1975. p. 287

[102] Igoni AH, Ayotamuno MJ, Eze CL, Ogaji SOT, Probert SD. Designs of anaerobic digesters for producing biogas from municipal solid-waste. Applied Energy. 2008;**85**:430-438

[103] Eckenfelder WW Jr. Industrial Water-Pollution Control. Boston Burr Ridge: McGraw-Hill Higher Education; 2000. pp. 394-411

[104] Ward AJ, Hobbs PJ, Holliman PJ, Jones DL. Optimisation of the anaerobic digestion of agricultural resources. Bioresource Technology. 2008;**99**:7928-7940

[105] Jain SR, Mattiasson B. Acclimatization of methanogenic consortia for low pH biomethanation process. Biotechnology Letters. 1998;**20**:771-775

[106] Choorit W, Wisarnwan P. Effect of temperature on the anaerobic digestion of palm oil mill effluent. Electronic Journal of Biotechnology. 2007;**10**:376-385

[107] Kumar S. Biogas. Rijeka, Croatia: Intech; 2012. Available from: www.intechopen.com/books/biogas

[108] Kahaynian M, Lindenauer K, Hardy S, Tchobanoglous G. Two-stage process combines anaerobic and aerobic methods. Biocycle. 1991;**32**(3): 48-53

[109] Kumar A, Miglani P, Gupta RK, Bhattacharya TK. Impact of Ni(II), Zn(II) and Cd(II) on biogassification of potato waste. Journal of Environmental Biology. 2006;**27**:61-66

[110] Oregon State Department of Energy. Biomass Energy Technology [Internet]. 2002. Available from: http://www.oregondoe.org/

[111] Tufaner F, Avşar Y. Effects of co-substrate on biogas production from cattle manure: A review. International Journal of Environmental Science and Technology. 2016;**13**:2303-2312

[112] Gomez-Lahoz C, Fernandez GB, Garcia HF, Rodriguez MJM, Vereda AC. Biomethanization of mixtures of fruits and vegetables solid wastes and sludge from a municipal waste water treatment plant. Journal of Environmental Science and Health. 2007;**42**:481-487

[113] Mshandete A, Bjornsson L, Kivaisi AK, Rubindamayugi MST, Mattiasson B. Effect of particle size on biogas yield from sisal fibre waste. Renewable Energy. 2006;**31**:2385-2392

[114] Sharma SK, Mishra IM, Sharma MP, Saini JS. Effect of particle size on biogas generation from biomass residues. Biomass. 1988;**17**:251-263

[115] Soccol RC et al. Lignocellulosic bioethanol: Current status and future perspectives. In: Biofuels. Amsterdam: Academic Press; 2011. pp. 101-122

[116] Ahring BK et al. In: Mata-Alvarez J, editor. Biomethanization of the Organic

Fraction of Municipal Solid Wastes. London, UK: IWA Publishing; 2003

[117] Korres N et al. Bioenergy Production by Anaerobic Digestion Using Agricultural Biomass and Organic Wastes. UK: Routledge Taylor and Francis Group; 2013

[118] Verma S. Anaerobic digestion of biodegradable organics in municipal solid wastes [M.Sc. dissertation]. USA: Department of Earth and Environmental Engineering, School of Engineering and Applied Science, Columbia University; 2002

[119] Wellinger A, Wyder K, Metzler AE. Kompogas—A new system for the anaerobic treatment of source separated waste. Water Science and Technology. 1993;**27**(2):153-158

[120] Luning L, van Zundert EH, Brinkmann AJ. Comparison of dry and wet digestion for solid waste. Water Science and Technology. 2003;**48**(4):15-20

[121] Malakahmad A, Ahmad Basri N, Md Zain S. An application of aerobic baffled reactor to produce biogas from kitchen waste. WIT Transactions on Ecology and the Environment at the Fourth International Conference on Waste Management and the Environment. 2008;**109**:655-664. DOI: 10.2495/WM080671

[122] Berglund M, Börjesson P. Assessment of energy performance in the life-cycle of biogas production. Biomass and Bioenergy. 2006;**30**:254-266

[123] Bouallagui H et al. Bioreactor performance in anaerobic digestion of fruit and vegetable wastes. Process Biochemistry. 2005;**40**(3):989-995

[124] Mathias JFC. Manure as a resource: Livestock waste management from anaerobic digestion, opportunities and challenges in Brazil. International Food and Agribusiness Management Review. 2014;**17**:87-110

[125] Browne JD, Allen E, Murphy JD. Improving hydrolysis of food waste in a leach bed reactor. Waste Management. 2013;**33**:2470-2477

[126] Carrillo-Reyes J, Cortés-Carmona MA, Bárcenas-Ruiz CD, Razo-Flores E. Cell wash-out enrichment increases the stability and performance of biohydrogen producing packed-bed reactors and the community transition along the operation time. Renewable Energy. 2016;**97**:266-273

[127] Kim M, Gomec CY, Ahn Y, Speece R. Hydrolysis and acidogenesis of particulate organic material in mesophilic and thermophilic anaerobic digestion. Environmental Technology. 2003;**24**:1183-1190

[128] Rajeshwari K, Balakrishnan M, Kansal A, Lata K, Kishore V. State-of-the-art of anaerobic digestion technology for industrial wastewater treatment. Renewable and Sustainable Energy Reviews. 2000;**4**:135-156

[129] Sharma V, Testa C, Lastella G, Cornacchia G, Comparato M. Inclined-plug-flow type reactor for anaerobic digestion of semi-solid waste. Applied Energy. 2000;**65**:173-185

[130] Mao C, Feng Y, Wang X, Ren G. Review on research achievements of biogas from anaerobic digestion. Renewable and Sustainable Energy Reviews. 2015;**45**:540-555

[131] Patinvoh RJ et al. Dry fermentation of manure with straw in continuous plug flow reactor: Reactor development and process stability at different loading rates. Bioresource Technology. 2017;**224**:197-205

[132] Capela I, Bilé MJ, Silva F, Nadais H, Prates A, Arroja L. Hydrodynamic

behaviour of a full-scale anaerobic contact reactor using residence time distribution technique. Journal of Chemical Technology and Biotechnology. 2009;**84**:716-724

[133] Şentürk E, İnce M, Engin GO. Kinetic evaluation and performance of a mesophilic anaerobic contact reactor treating medium-strength food-processing wastewater. Bioresource Technology. 2010;**101**: 3970-3977

[134] Şentürk E, İnce M, Engin GO. The effect of transient loading on the performance of a mesophilic anaerobic contact reactor at constant feed strength. Journal of Biotechnology. 2013;**164**:232-237

[135] Aslanzadeh S, Rajendran K, Jeihanipour A, Taherzadeh MJ. The effect of effluent recirculation in a semi-continuous two-stage anaerobic digestion system. Energies. 2013;**6**:2966-2981

[136] Lo K, Liao P. Digestion of cheese whey with anaerobic rotating biological contact reactors. Biomass. 1986;**10**:243-252

[137] Wei C-H, Harb M, Amy G, Hong P-Y, Leiknes T. Sustainable organic loading rate and energy recovery potential of mesophilic anaerobic membrane bioreactor for municipal wastewater treatment. Bioresource Technology. 2014;**166**:326-334

[138] Chen C, Guo W, Ngo HH, Lee DJ, Tung KL, Jin P, et al. Challenges in biogas production from anaerobic membrane bioreactors. Renewable Energy. 2016;**98**:120-134

[139] Karadag D, Köroğlu OE, Ozkaya B, Cakmakci M. A review on anaerobic biofilm reactors for the treatment of dairy industry wastewater. Process Biochemistry. 2015;**50**:262-271

[140] Patel P, Desai M, Madamwar D. Biomethanation of cheese whey using anaerobic upflow fixed film reactor. Journal of Fermentation and Bioengineering. 1995;**79**:398-399

[141] Liu Y, Zhu Y, Jia H, Yong X, Zhang L, Zhou J, et al. Effects of different biofilm carriers on biogas production during anaerobic digestion of corn straw. Bioresource Technology. 2017;**244**:445-451

[142] Khalid A et al. The anaerobic digestion of solid organic waste. Waste Management. 2011;**31**(8):1737-1744

[143] Sharma VK, Fortuna F, Canditelli M, Cornacchia G, Farina R. Anaerobic digester for treatment of organic waste. In: ENEA (Energia Nucleare e delle Energie Alternative)-RT-AMB-97-17. Rome (Italy): Technical Report; 1997. pp. 1-47

[144] Chiumenti R, De Poli F, Gamboni M, Gruercini S, Renalli G, Sorlini C. Anaerobic digestion of liquid fraction of beef cattle waste in an advanced hybrid bioreactor. In: Neijssel OM et al., editors. Proceedings of the 4th European Congress on Biotechnology. Amsterdam; 1987. pp. 204-207

[145] De Poli F, Farina R, Giamboni M, Garuti G, Tilche A. ENEA's activities on advanced anaerobic treatment processes. Rome, Italy: ENEA; 1988

[146] Ke S, Shi Z, Fang HH. Applications of two-phase anaerobic degradation in industrial wastewater treatment. International Journal of Environment and Pollution. 2005;**23**:65-80

[147] Lehtomäki A, Huttunen S, Lehtinen T, Rintala J. Anaerobic digestion of grass silage in batch leach bed processes for methane production. Bioresource Technology. 2008;**99**:3267-3278

[148] Budzianowski WM. A review of potential innovations for production, conditioning and utilization of biogas with multiple-criteria assessment. Renewable and Sustainable Energy Reviews. 2016;**54**:1148-1171

[149] Nair S, Kuang Y, Pullammanappallil P. Enhanced degradation of waste grass clippings in one and two stage anaerobic systems. Environmental Technology. 2005;**26**:1003-1012

[150] Ahamed A, Chen C-L, Rajagopal R, Wu D, Mao Y, Ho I, et al. Multi-phased anaerobic baffled reactor treating food waste. Bioresource Technology. 2015;**182**:239-244

[151] Vandevivere P, Baere LD, Verstraete W. Types of anaerobic digester for solid wastes. In: Mata-Alvarez J, editor. Biomethanization of the Organic Fraction of Municipal Solid Wastes. London: IWA Publishing; 2003. pp. 111-140

[152] Kelleher M. Anaerobic digestion outlook for MSW streams. Biocycle. 2007;**48**:51

[153] Ghosh S et al. Pilot-scale gasification of municipal solid wastes by high-rate and two-phase anaerobic digestion (TPAD). Water Science and Technology. 2000;**41**(3):101-110

[154] Ince O. Performance of a two-phase anaerobic digestion system when treating dairy wastewater. Water Research. 1998;**32**(9):2707-2713

[155] Chaudhary BK. Dry continuous anaerobic digestion of municipal solid waste in thermophilic conditions [Master of Engineering thesis]. Thailand: School of Environment, Resources and Development, Asian Institute of Technology; 2008. Available from: http://faculty.ait.ac.th/visu/public/uploads/Data/AITThesis/Master%20Thesis%20final/Binod.pdf

[156] Joshua LR et al. Anaerobic digestion technologies for the treatment of municipal solid waste. International Journal of Environment and Waste Management. 2012;**9**(1-2, 122):100

[157] Griffin LP. Anaerobic digestion of organic wastes: The impact of operating conditions on hydrolysis efficiency and microbial community composition [Master of Science thesis]. Fort Collins, Colorado: Department of Civil and Environmental Engineering, Colorado State University; 2012

[158] Petersson A, Wellinger A. Biogas upgrading technologies - Developments and innovations. Task 37: Energy from biogas and Landfill Gas. Paris, France: IEA Bioenergy; 2006. Available from: https://www.ieabioenergy.com/publications/biogas-upgrading-technologies-developments-and-innovations/

[159] Vijay VK. Biogas Enrichment and Bottling Technology for Automobile Fuel-IIT Delhi Technology - Case Study of Goshala in Rajasthan. New Delhi, India: Centre for Rural Development & Technology, Indian Institute of Technology Delhi; 2012

[160] Abatzoglou N, Boivin S. A review of biogas purification processes. Biofuels, Bioproducts and Biorefining. 2009;**3**:42-71

[161] Nickel K, Neis U. Ultrasonic disintegration of biosolids for improved biodegradation. Ultrasonics Sonochemistry. 2007;**14**:450-455. DOI: 10.1016/j.ultsonch.2006.10.012

[162] Xu J, Yuan H, Lin J. Evaluation of thermal, thermal-alkaline, alkaline and electrochemical pretreatments on sludge to enhance anaerobic biogas production. Journal of the Taiwan Institute of

Chemical Engineers. 2014;**45**:2531-2536. DOI: 10.1016/j.jtice.2014.05.029

[163] Farghali M, Andriamanohiarisoamanana FJ, Ahmed MM, et al. Prospects for biogas production and H_2S control from the anaerobic digestion of cattle manure: The influence of microscale waste iron powder and iron oxide nanoparticles. Waste Management. 2020;**101**:141-149. DOI: 10.1016/j.wasman.2019.10.003

[164] Mrosso R, Machunda R, et al. Removal of hydrogen sulfide from biogas using a red rock. Journal of Energy. 2020;**2020**:1-10. ID: 2309378. DOI: 10.1155/2020/2309378

[165] Curto D, Martín M. Renewable based biogas upgrading. Journal of Cleaner Production. 2019;**224**:50-59

[166] Kougias PG, Treu L, Benavente DP, Boe K, Campanaro S, Angelidaki I. Ex-situ biogas upgrading and enhancement in different reactor systems. Bioresource Technology. 2017b;**225**:429-437

[167] Eze JI, Agbo KE. Maximizing the potentials of biogas through upgrading. American Journal of Scientific and Industrial Research. 2010;**1**:604-649

[168] Bharathiraja B, Sudharsana T, Jayamuthunagai J, Praveenkumar R, Chozhavendhan S, Iyyappan S. Biogas production—A review on composition, fuel properties, feed stock and principles of anaerobic digestion. Renewable and Sustainable Energy Reviews. 2018;**90**:570-582

[169] Hoogendoorn A, van Kasteren H. Transportation Biofuels: Novel Pathways for the Production of Ethanol, Biogas and Biodiesel. Cambridge: Royal Society of Chemistry; 2010. p. 202

[170] Thrän D, Billig E, Persson T, Svensson M, Daniel-Gromke J, Ponitka J, Seiffert M. Biomethane—Status and Factors Affecting Market Development and Trade. IEA Task 40 and Task 37 Joint Study. 2014. Available from: http://task40.ieabioenergy.com/wpcontent/uploads/2013/09/t40-t37-biomethane-2014.pdf [Accessed: 18 January 2018]

[171] Augelletti R, Conti M, Annesini MC. Pressure swing adsorption for biogas upgrading. A new process configuration for the separation of biomethane and carbon dioxide. Journal of Cleaner Production. 2017;**140**:1390-1398

[172] Bauer F, Hulteberg C, Persson T, Tamm D. Biogas Upgrading - Review of Commercial Technologies. Sweden: Svenskt Gastekniskt Center AB (SGC Rapport). Vol. 270. 2013a

[173] Yousef AM, El-Maghlany WM, Eldrainy YA, Attia A. New approach for biogas purification using cryogenic separation and distillation process for CO_2 capture. Energy. 2018;**156**:328-351. DOI: 10.1016/j.energy.2018.05.106

[174] Tan Y, Nookuea W, Li H, et al. Cryogenic technology for biogas upgrading combined with carbon capture - A review of systems and property impacts. Energy Procedia. 2017b;**142**:3741-3746

[175] Grande CA, Blom R. Cryogenic adsorption of methane and carbon dioxide on zeolites 4A and 13X. Energy & Fuels. 2014;**28**:6688-6693. DOI: 10.1021/ef501814x

[176] Song C, Liu Q, Deng S, et al. Cryogenic-based CO_2 capture technologies: State-of-the-art developments and current challenges. Renewable and Sustainable Energy Reviews. 2019;**101**:265-278. DOI: 10.1016/j.rser.2018.11.018

[177] Riva M, Campestrini M, Toubassy J, et al. Solid-liquid-vapor

equilibrium models for cryogenic biogas upgrading. Industrial and Engineering Chemistry Research. 2014;**53**:17506-17514. DOI: 10.1021/ie502957x

[178] Awe OW, Zhao Y, Nzihou A, et al. A review of biogas utilisation, purification and upgrading technologies. Waste and Biomass Valorization. 2017;**8**:267-283. DOI: 10.1007/s12649-016-9826-4

[179] Niemczewska J. Characteristics of utilization of biogas technology. Nafta-Gaz. 2012;**68**:293-297

[180] Jürgensen L, Ehimen EA, Born J, Holm-Nielsen JB. Dynamic biogas upgrading based on the Sabatier process: Thermodynamic and dynamic process simulation. Bioresource Technology. 2015;**178**:323-329

[181] Xia A, Cheng J, Murphy JD. Innovation in biological production and upgrading of methane and hydrogen for use as gaseous transport biofuel. Biotechnology Advances. 2016;**34**:451-472

[182] Jürgensen L, Ehimen EA, Born J, Holm-Nielsen JB. Utilization of surplus electricity from wind power for dynamic biogas upgrading: Northern Germany case study. Biomass and Bioenergy. 2014;**66**:126-132

[183] Guebitz GM, Bauer A, Bochmann G, Gronauer A, Weiss S, editors. Biogas Science and Technology. Hannover, Germany: Springer; 2015

[184] Damyanova S, Beschkov V. Biogas as a source of energy and chemicals. In: Biorefinery Concepts. Rijeka, Croatia: Intech; 2019. DOI: 10.5772/intechopen.90558. [Accessed: 19 July 2020]

[185] Harasimowicz M, Orluk P, Zakrzewska TG, Chmielewski AG. Application of polyimide membranes for biogas purification and enrichment. Journal of Hazardous Materials. 2007;**144**:698-702

[186] Kapdi SS, Vijay VK, Rajesh SK, Prasad R. Biogas scrubbing, compression and storage: Perspective and prospectus in Indian context, renew. Energy. 2005;**30**(8):1195-1202

[187] Appels L, Joost L, Degreve J, Helsen L, Lievens B, Willems K, et al. Anaerobic digestion in global bio-energy production: Potential and research changes. Renewable and Sustainable Energy Reviews. 2011;**15**:4295-4301

[188] World Energy Council. World Energy Resources: Waste to Energy. World Energy Council; 2013. Available from: https://www.worldenergy.org/wp-content/uploads/2013/10/WER_2013_7b_Waste_to_Energy.pdf

[189] Adnan AI, Ong MY, Nomanbhay S, Chew KW, Show PL. Technologies for biogas upgrading to biomethane: A review. Bioengineering. 2019;**6**:92

[190] Teixeira Coelho S, Stortini Gonzales Velazques SM, Stella Martins O, Castro De Abreu F. Biogas from sewage treatment used to electric energy generation, by a 30 kW (ISO) microturbine. In: Proceedings of the World Bioenergy Conference & Exhibition; 30 May-1 June 2006; Jönköping, Sweden. Available from: http://143.107.4.241/download/projetos/2_Erg-bior.pdf

[191] Schaller M. Biogas electricity production hits 17,272 GWh a year in Europe. In: Engineer Live, London. 2007. pp. 46-49. Available from: https://sswm.info/node/503 [Accessed: 20 July 2020]

[192] Sarkar SC, Bose A. Role of activated carbon pellets in carbon dioxide removal. Energy Conversion and Management. 1997;**38**(Suppl 1): S105-S110

[193] O'Hayre RP, Cha S-W, Colella WG, Prinz FB. Fuel Cell Fundamentals. 2nd ed. New Jersey, USA: John Wiley and Sons, Inc.; 2009. pp. 384-385

[194] Biogas for Road Vehicles. Technology Brief. Available from: https://www.irena.org/-/media/Files/IRENA/Agency/Publication/2017/Mar/IRENA_Biogas_for_Road_Vehicles_2017.pdf [Accessed: 13 November 2019]

[195] Jönsson O, Persson M. Biogas as transportation fuel. FVS Fachtagung. 2003;**99**:99-111

[196] Schnurer A, Jarvis A. Microbiological Handbook for Biogas Plants. Sweden: Swedish Waste Management, Swedish Gas Centre, Malmö; U2009;**03**:1-74

[197] Finnerty WR. Microbial lipid metabolism. In: Microbial Lipids. Vol. 2. San Diego: Academic Press; 1989. pp. 525-566

[198] Wackett LP, Frias JA, Seffernick JL, Sukovich DJ, Cameron SM. Genomic and biochemical studies demonstrating the absence of an alkane-producing phenotype in *Vibrio furnissii* M1. Applied and Environmental Microbiology. 2007;**73**:7192-7198

[199] Gopalakrishnan H, van Leeuwen J, Brown R. Sustainable Bioenergy and Bioproducts. Value Added Engineering and Applications. London: Springer; 2012

[200] Jin B, Tin P, Ma Y, Zhao L. Production of lactic acid and fungal biomass by *Rhizopus* fungi from food processing waste streams. Journal of Industrial Microbiology & Biotechnology. 2005;**32**:678-686

[201] Ngô C. Energy: Resources, Technologies and the Environment. London: The Institution of Engineering and Technology; 2010

[202] Shivali S, Goldy S, Pooja G, Rimika K, Subhanjan S, Priyanka S, et al. Review of trends in biogas upgradation technologies and future perspectives. Bioresource Technology. 2018;**1**:79-88

[203] Martins das Neves LC, Converti A, Vessoni Penna TC. Biogas production: New trends for alternative energy sources in rural and urban zones. Chemical Engineering and Technology. 2009;**32**:1147-1153

[204] Yu L, Yaoqiu K, Ningsheng H, Zhifeng W, Lianzhong X. Popularizing household-scale biogas digesters for rural sustainable energy development and greenhouse gas mitigation. Renewable Energy. 2008;**33**:2027-2035

[205] Lakhal S, H'Mida S. A gap analysis for green supply chain benchmarking. In: 32th International Conference on Computers & Industrial Engineering. Vol. 1; Ireland; 11-13 August 2003. pp. 44-49

[206] Islam MR, Chhetri AB, Khan MM. Green Petroleum: How Oil and Gas Can Be Environmentally Sustainable. Do You Believe in Global Warming?. US: Wiley; 2012. p. 371

[207] Raboni M, Urbini G. Production and use of biogas in Europe: A survey of current status and perspectives. Ambiente & Água. 2014;**9**:191-202. DOI: 10.4136/ambi-agua.1324

[208] FAO. Biofuels: Prospects and Opportunities. Rome: FAO; 2008

[209] Janssen R, Rutz D, editors. Bioenergy for Sustainable Development in Africa. Dordrecht/Heidelberg/London/New York: Springer; 2012

[210] Dahlgren S. The role of biogas in a more sustainable energy system in Sweden [licentiate thesis]. Sweden: Linköping University Electronic Press; 2019

[211] Iglinski B, Buczkowski R, Iglinska A, Cichosz M, Piechota G, Kujawski W. Agricultural biogas plants in Poland: Investment process, economical and environmental aspects, biogas potential. Renewable and Sustainable Energy Reviews. 2012;**16**:4890-4900

[212] Curkowski A, Oniszk A, Mroczkowski P, Owsik M, Wis GA. Guide for Investors Interested in Construction of Agricultural Biogas Plants. Warsaw: Institute for Renewable Energy; 2011

[213] Garfi M, Ferrer L, Velo E, Ferrer I. Evaluating benefits of low-cost household digestors for rural Anden communities. Renewable and Sustainable Energy Reviews. 2012;**16**:574-581

[214] Bond T, Templeton MR. History and future of domestic biogas plants in the developing world. Energy for Sustainable Development. 2011;**15**:347-354

[215] Tomassetti L, Segreto M, Torre M, Borin D, Tratzi P, Paolini V, et al. Environmental impact of biogas in Europe. In: Biomass Sustainability, Impacts and Policies; 27th European Biomass Conference and Exhibition; 27-30 May 2019; Lisbon, Portugal. 2019

[216] Massé DI, Talbot G, Gilbert Y. On farm biogas production: A method to reduce GHG emissions and develop more sustainable livestock operations. Animal Feed Science and Technology. 2011;**166-167**:436-445

[217] Pathak H, Jain N, Bhatia A, Mohanty S, Gupta N. Global warming mitigation potential of biogas plants in India. Environmental Monitoring and Assessment. 2009;**157**:407-418

[218] Winquist E, Rikkonen P, Pyysiäinen J, Varho V. Is biogas an energy or a sustainability product ? - Business opportunities in the Finnish biogas branch. Journal of Cleaner Production. 2019;**233**:1344-1354. DOI: 10.1016/j.jclepro.2019.06.181

[219] Bhardwaj S, Das P. A review: Advantages and disadvantages of biogas. International Research Journal of Engineering and Technology. 2017;**4**(10):890-893

[220] Callaghan FJ, Wase DAJ, Thayanithy K, Foster CF. Co-digestion of waste organic solids: Batch studies. Bioresource Technology. 1999;**67**:117-122

[221] Agdag ON, Sponza DT. Co-digestion of mixed industrial sludge with municipal solid wastes in anaerobic simulated landfilling bioreactors. Journal of Hazardous Materials. 2007;**140**:75-85

Chapter 2

Dry Anaerobic Digestion for Agricultural Waste Recycling

Shohei Riya, Lingyu Meng, Yuexi Wang, Chol Gyu Lee, Sheng Zhou, Koki Toyota and Masaaki Hosomi

Abstract

For sustainable agriculture, it is important to manage agricultural wastes, such as crop residues and livestock wastes. Anaerobic digestion has been gathering the attention to recycle these wastes into renewable energy (biogas) and fertilizer (soil amendment) (digestate). Dry anaerobic digestion is defined as digestion at higher than 20% of total solid (TS) content in the reactor, which is suitable for wastes with high TS content, such as agricultural wastes. In this chapter, we reviewed recent advances in biogas production and use of digestate as soil amendment from dry anaerobic digestion of agricultural wastes. It has been found that ammonia concentration, feed/inoculum (F/I) ratio, and TS content are important parameters for operation of dry anaerobic digestion. Several operation technologies have been in operation, while new operation strategies have been developed. Application of solid digestate into the soil is beneficial to increase soil properties; however it should be carefully operated because it has risks of nitrate leaching and soil pathogens.

Keywords: dry anaerobic digestion, biogas, crop residue, manure, soil amendment

1. Introduction

Providing energy and food with low environmental impact is considered as an urgent issue in order to meet demands of them for the growing global population. Alternative resources to replace fossil fuel for energy and chemical fertilizer production are required. Agricultural wastes, such as crop residues and livestock wastes, have been gathering attention as a source of renewable energy and nutrient [1]. Agricultural waste such as lignocellulosic biomass is available globally over 200 billion dry metric ton per year [2]. Livestock wastes such as manures are important nutrient source. Global estimates of nitrogen and phosphorus in the manures were 128 and 24 Tg for 2007, which are almost two times higher than those of fertilized chemical fertilizer [3].

Anaerobic digestion is a technology for treatment of organic wastes, which can biologically decompose carbohydrates, proteins, and lipids in the absence of oxygen and produce biogas (CH_4 and CO_2). In anaerobic digestion, nitrogen in protein and amino acids are mineralized and transformed into ammonium (NH_4^+). Total P and K are also not lost and retained in the digestate [4]. These nutrients are retained in the residue of anaerobic digestion, called digestate. Therefore, anaerobic digestion can produce both renewable energy and nutrients. In addition to organic waste treatment, anaerobic digestion can be utilized for effective biological pretreatment for anaerobic biorefinery [5]. In the anaerobic biorefinery concept, biogas is further transformed into alcohol or syngas, etc., and digestate is utilized for algae, organic acid, and alcohol biopolymer productions [6]. Digestate can be also applied to agricultural land as a fertilizer [7] for production of crops or forages since it contains nutrients as noted above. Recycling digestate as a fertilizer can reduce chemical fertilizer production, hence reducing fossil fuel consumption and CO_2 emission [8]. Harvested crop residues and collected manures from the livestock fed with the harvested forage can be used for substrate in anaerobic digestion. Thus, anaerobic digestion can be a key technology to recycle waste into value-added products and fertilizer.

Generally, anaerobic digestion is conducted in the form of liquid at low total solid (TS) content less than 15% [9], called wet anaerobic digestion. Wet anaerobic digestion is suitable for wastes with low TS contents (high-moisture contents) [10]. However, to maintain low TS content in the reactor, it requires a large amount of water if it treats wastes with high TS content, such as lignocellulosic biomass, resulting in increase in reactor volume as well as generation of a huge volume of wastewater to be treated [9]. In addition, digested slurry is subjected to solid–liquid separation process [11] after wet anaerobic digestion for further processing.

On the contrary to wet anaerobic digestion, operation at TS content of higher than 15% is classified as dry (solid-state) anaerobic digestion [9]. Dry anaerobic digestion has several advantages over wet anaerobic digestion such as less fresh water usage and favorable energy balance [12]. Agricultural waste such as lignocellulosic biomass has high TS content. For example, TS contents of the corn silage, grasses, and straw biomasses are 25–89% [13]. For livestock manure, depending on pretreatment (solid–liquid separation), TS contents of solid phase are 18–30% [13, 14]. Therefore, agricultural wastes are suitable in dry anaerobic digestion in terms of TS content. Total solid contents of the solid fraction after solid–liquid separation of wet digestate are 23–30% [15], which are comparable or slightly higher than TS content of the dry anaerobic digestate (TS content in the reactor) [16]. Therefore, it would be expected that dry anaerobic digestion would reduce post-digestate treatment such as solid–liquid separation and treatment of liquid fraction, which can reduce energy consumption and cost for plant construction and operation. Therefore, dry anaerobic digestion would have more advantages over wet anaerobic digestion for biorefinery of agricultural wastes.

Although dry anaerobic digestion has several benefits, still wet anaerobic digestion plants have more advantages in terms of energy balance and cost performance in practice [12], requiring more research on effective operation of dry anaerobic digestion. Operation parameters of dry anaerobic digestion should be carefully determined. In general, mass transfer in the dry digestion media is not adequate, and high organic loading would reduce degradation of substrate and biogas production [10]. In addition, treatment of waste with high nitrogen concentration, such as manure, would result in ammonia accumulation and failure [17].

Digestate from the anaerobic digester can be used as fertilizer as it contains nutrient for crop growth or further processed to produce value-added products as noted above. For digestate from wet anaerobic digestion, digestates are subjected to solid–liquid separation [18]. These liquid fraction and solid fraction can be used as fertilizer [18]. Numerous studies have been conducted to evaluate effect of digestate from the wet anaerobic digestion on crop production and environmental risks [15], while digestate from the dry anaerobic digestion has not been well studied.

In this chapter, we reviewed research progress in dry anaerobic digestion of agricultural waste. The key parameters and reactor types of dry anaerobic digestion were summarized. In terms of digestate recycling, we focused on the application of digestate in agricultural land. Especially, the effect of digestate from the wet and dry anaerobic digestion on soil nitrate leaching and root-knot nematodes was summarized.

2. Key parameters of dry anaerobic digestion

Anaerobic digestion is conducted by anaerobic microorganisms contributing to hydrolysis, acid production, and methane production. Therefore, operation parameters should be taken into account for their growth and inhibition. For example, manures containing high concentration of ammonia causes ammonia inhibition. In addition, higher TS content in the dry anaerobic digester causes slow mass transfer, resulting in slow decomposition of intermediate. The accumulation of the intermediate will result in inhibition of methane production. In this section, important parameters of dry anaerobic digestion were reviewed.

2.1 Ammonia concentration

Nitrogen is an essential nutrient for microorganisms conducting anaerobic digestion. However, excess amount of nitrogen causes inhibition. According to Rajagopal et al. [19], ammonia concentration between 50 and 200 mg N L^{-1} is beneficial for anaerobic digestion while higher than 1500 mg N L^{-1} inhibits digestion. In the solution, ammonium ion (NH_4^+) is equilibrated with free ammonia (NH_3). The equilibrium is governed by pH and temperature [20]. Therefore, higher pH and higher temperature increase NH_3 concentration. Free ammonia can diffuse into the cell through the cell membrane and inhibits cell function by disrupting the proton and potassium balance [21]. Therefore, thermophilic (55°C) condition is more sensitive to ammonia inhibition than mesophilic (37°C) condition.

In the dry anaerobic digestion, ammonia inhibition was reported in digestion of high nitrogen-containing biomass or digestion of low nitrogen-containing biomass with inoculum with high nitrogen concentration. Under thermophilic conditions, dry anaerobic digestion of corn stover highly inoculated with wet anaerobic digestion effluent showed smaller amount of biogas production than those with less inoculated one [22]. This was due to high concentration of ammonia in the inoculum. In semi-solid (10% of TS) digestion of chicken manure, 12 and 16 g N L^{-1} of ammonia were accumulated in mesophilic and thermophilic conditions, respectively, and the mesophilic condition showed higher methane production than that of thermophilic one [23]. Zhou et al. also observed low methane yield of thermophilic anaerobic digestion of pig manure, in which NH_4^+ concentration exceeded 4000 mg N kg^{-1} [17].

In order to overcome ammonia inhibition, several approaches were suggested such as ammonia striping, chemical precipitation, adjusting carbon/nitrogen (C/N) ratio, etc. Ammonia stripping was applied for dry anaerobic digestion of chicken manure. Ammonia in the chicken manure was stripped at high pH with N_2 flow after ammonia production by anaerobic fermentation [24]. Ammonia-stripped chicken manure showed 2305 mL kg^{-1} TS of cumulative methane production, which is much higher than the manure without stripping (313 mL kg^{-1} TS) [24]. In anaerobic digestion, C/N ratio of 15–30 is thought to be preferable [25]. A simple way to avoid ammonia inhibition is co-digestion with biomass with low nitrogen content such as crop residue. Co-digestion can dilute ammonia concentration in the reactor and reduce ammonia inhibition. For example, Abouelenien et al. found 1.5–93% increase in methane production in thermophilic co-digestion of chicken manure (C/N ratio of 6) with agricultural waste (coconut, coffee grounds, and cassava; C/N ratio of 17) compared with mono-digestion of chicken manure. Zhou et al. mixed pig manure with rice straw to obtain mixtures with C/N ratio of 10, 20, and 30 and conducted dry thermophilic digestion. The methane yields of C/N ratio of 20 and 30 were 244 and 258 mL g^{-1} VS, while C/N ratio of 10 showed lower and unstable methane production [17].

2.2 F/I ratio

In batch dry anaerobic digestion, the ratio of feed (substrate) and inoculum (F/I or S/I ratio), which is an index of organic loading to microorganisms, is an important parameter for efficient digestion. Operation with higher F/I ratio can treat larger amount of substrate in one batch. In the studies of dry anaerobic digestion, F/I ratios of 0.5–10 were applied or evaluated [22, 26–28]. Generally, increase in F/I ratio results in slower startup and lower methane yield than those of lower F/I ratio. For example, in mesophilic dry anaerobic digestion of corn stover, F/I ratio of 2.43 showed the highest methane yield (321 L kg^{-1}), followed by F/I ratios of 3.44, 4.58, and 7.41 [22]. Co-digestion of rape straw and dairy manure also showed higher methane yield (209 L kg^{-1}) in low F/I ratio (2:3 of feed/inoculum) than those in higher F/I ratio [29].

The reason why dry digestion at higher F/I ratio failed is acidification of the reactors by accumulation of volatile organic acids (VFAs). Li et al. observed that the final pH in failed reactors at F/I ratio of 4.58 and 7.41 were 5.43 and 5.11, respectively, in digestion of corn stover [22]. According to the VFA concentration and pH changes during digestion, overaccumulation of VFAs (up to 25 g L^{-1}) and drop of pH (less than 6) caused inhibition of methane production at high F/I ratio (3 and 4) [29]. Most methanogens are active in pH of 6.6–7.6 with an optimum pH of ca. 7. Therefore, acidification by accumulation of VFAs causes reduction of methane production activity. In addition to its influence to methanogens, high F/I ratio affects hydrolysis. Cui et al. observed cellulose and hemicellulose degradation rates were about 40% in dry anaerobic digestion of spent wheat straw at F/I ratio of 2 and 4, while it was less than 10% at F/I ratio of 6 [30]. Similar results were also observed in dry anaerobic digestion of solid waste residues of palm oil mill industry [31]. At pH of 6, the performance of hydrolysis and VFAs production was lower than at higher pH in fermentation of lignocellulosic waste [32]. Therefore, lowering pH may affect all the processes of anaerobic digestion (hydrolysis, VFA production, and methane production).

2.3 Total solid content

High TS content can reduce reactor volume and capital cost [9]. However, in dry anaerobic digestion, higher TS content reduces methane production. Xu et al. reported that maximum methane production rates were proportionally increased with TS content between 0 and 20% while gradually decreased from 20% TS to 30% TS content in mesophilic digestion of corn stover [33]. For mesophilic dry digestion of empty fruit bunch and oil palm trunk, methane yields at 16 and 25% TS contents were 250–350 mL g^{-1} VS. At 35% TS content, however methane yields were less than 100 mL g^{-1} VS with some exception [31]. In semi-batch dry thermophilic co-digestion of pig manure and rice straw, biogas yields were around 600 mL g^{-1} VS, and no VFAs accumulation was observed between 18% and 27% of TS content in the reactor [16]. However, biogas production was decreased concomitantly with VFAs accumulation when TS content in the reactor exceeded 28% [16]. Therefore, TS content should be carefully chosen and managed.

According to Le Hyaric et al., increasing TS content resulted in linear decrease in methane production from acetate, propionate, and cellulose [34]. They pointed out that acetate removal is a rate-limiting step in dry anaerobic digestion since H_2 produced from cellulose degradation was rapidly consumed and showed higher methane production than degradation of acetate [34]. However, there have been less information on rate-limiting step at high TS content. More study is required.

It has been thought that slow solute transport would cause reduction of biogas production at high TS content. In the dry anaerobic digestion, molecular diffusion is thought to control solute transport within the digestion medium because mixing is limited [35]. Solute flux by molecular diffusion is proportional to solute concentration gradient. And its proportional constant, called diffusion coefficient, characterizes the extent of the solute transport by molecular diffusion. Less information are available on the measurement of diffusion coefficient in the dry anaerobic digestion media. Several studies measured diffusion coefficient at high TS content. According to Bollon et al., diffusion coefficient of the solutes in the water is in the order of 10^{-9} m^2 s^{-1} while in the order of 10–11 m^2 s^{-1} at 8–25% of TS in digestate of the biowaste [35]. Similar results were also obtained by Zhang et al., who measured dewatered and digested sludge at 6–15% of TS content. Abbassi-Guendouz et al. demonstrated that limiting the overall mass transfer resulted in lower cumulative methane production [36].

3. Operating strategies of dry anaerobic digestion process

In dry anaerobic digestion process, major drawbacks are the heterogeneous distribution of substrate and microorganisms as well as low mass transfer under high solid content (> 20%). Inoculation efficiency of substrate is reduced by these factors, which results in unstable operation and low methane yield [37, 38]. Thus, keeping the inoculating efficiency is a main challenge for the operation of dry anaerobic digestion process.

Over the past 30 years, dry anaerobic digestion process has been developed and marketed by different companies in Europe. Commercial dry anaerobic digestion processes such as Valorga, Dranco, Kompogas, Bekon, and Bioferm are the most prevalent processes for treating municipal solid waste (MSW), biowaste, livestock waste, as well as green waste (**Table 1**) [10, 39]. According to several reviews [39–41],

Technology			Waste	Temperature (°C)	TS (%)	SRT*/ Digestion period (days)	Biogas yield (m^3/t**)	Capacity (1000 tons/year)	Plants***
Full scale	Continuous	Valorga	MSW****	35, 55	25–35	16–35	80–160	10–498	27
		Dranco	MSW	55	20–50	13–30	103–147	3–320	32
		Kompogas	MSW, green waste	55	23–28	15–20	110–130	15–274	25
	Batch	Bekon	Biowaste, agricultural waste	35, 55	Na	28–35	130	4.5–60	60
		Bioferm	Food waste, green waste, agricultural waste	35	25	28	Na	8	9
							Methane yield	Scale (L)	
New case studies	Continuous	Kim and Oh [49]	Food waste, livestock waste	35	30–50	30–100	250 L/g COD	60	
		Zeshan et al. [48]	MSW	35–55	18	13–153	121–327 L/kg VS	690	Na
	Batch process	Meng et al. [51]	Rice straw, pig manure	55	20	40	191 L/kg VS	0.5	

*: Sludge retention time
**: Wet weight base
***: Accessed at 30 December 30, 2019
****: Municipal solid waste
Na: No data

Table 1.
Performance and parameters of commercial and new case studies of dry anaerobic digestion process. source: Data from the company websites as of December 2019 and adapted from Nichols [45], Lei et al. [40] and Andre et al. [39].

current strategies for improving the inoculating efficiency in dry anaerobic digestion process are mainly based on two considerations: (1) to homogenize the distribution of substrate and microorganisms by mechanical (biogas) mixing and (2) to improve the mass transfer in digester by the recirculation of liquid digestate. Also, some new efforts for improving the performance of dry anaerobic digestion process also have been conducted.

3.1 Homogenization

To improve homogenization, several different types of continuous dry anaerobic digestion processes such as Valorga (France), Kompogas (Switzerland), and Dranco (Belgium) have been proposed. In continuous digesters, wastes (substrate) are added to the digester at regular intervals, and equal amounts of finished products (digestate) are removed. For example, Valorga process sets a central baffle in the vertical steel tank, and the baffle extends two thirds of the way through the center of the tank. Wastes are forced to flow around the baffle from the inlet to reach the outlet port on the opposite side, creating a plug flow in the reactor. Pressured biogas is provided at the base of the tank at intervals, which promotes the moving up of wastes to the opposite side of the tank and the contact between wastes and mature digestate (**Figure 1a**). This process was operated under the following conditions: total solid content of 25–35% and sludge retention time (SRT) of 15–20 days. Approximately 80–160 m^3 t^{-1} of biogas can be recovered [42, 43]. Solid digestate generated from the process can be used as soil amendment after being dewatered and stored under aerobic conditions [40].

Similar to Valorga process, vertical tank is also used in Dranco process. However, different to Valorga process, Dranco process performs the mixing of wastes and finished digestates by a special pump (mix and introduce the mixture of wastes and finished digestates to the pipeline) before introducing the mixture into the inlet located at the top of the tank. Thereafter, introduced mixture moves from the top to the bottom (outlet) by gravity without any internal mixing mechanism during digestion (**Figure 1b**). Total solid content in Dranco process usually ranges from 20 to 50%, while the SRT ranges from 13 days to 30 days. Approximately 103–147 m^3 t^{-1} of biogas can be recovered [41, 44].

Different to Valorga and Dranco processes, Kompogas digester is a horizontal steel tank with slowly rotating axial mixers that assist in conveying the material from the inlet to the outlet, keep heavy solids in suspension, and degas the thick digestate. Finished digestates are recycled to inoculate the fresh wastes (**Figure 1c**). TS in Kompogas process usually ranges from 23 to 28%, and processed water may be added to reduce the solid content, while the SRT ranges from 15 days to 20 days. Approximately 110–130 m^3 t^{-1} of biogas can be recovered [41, 45].

3.2 Promotion of mass transfer

In order to improve the mass transfer in the digester, the batch dry anaerobic digestion process with percolation system has been proposed. This system recycles leachate into the digester and enables the colonization of bacteria throughout the digester by promoting the transport of microbes and dissolved substrate. Premix of wastes and finished digestate is usually performed to inoculate the wastes. Currently, Bekon (Germany) has the main market share in batch dry anaerobic digestion process. As shown in the diagram of Bekon process (**Figure 1d**) [46], the premixed wastes and finished digestate are set in the "garage-type" digester, and leachate is collected from the bottom of the digester (digester at a 15 degree angle

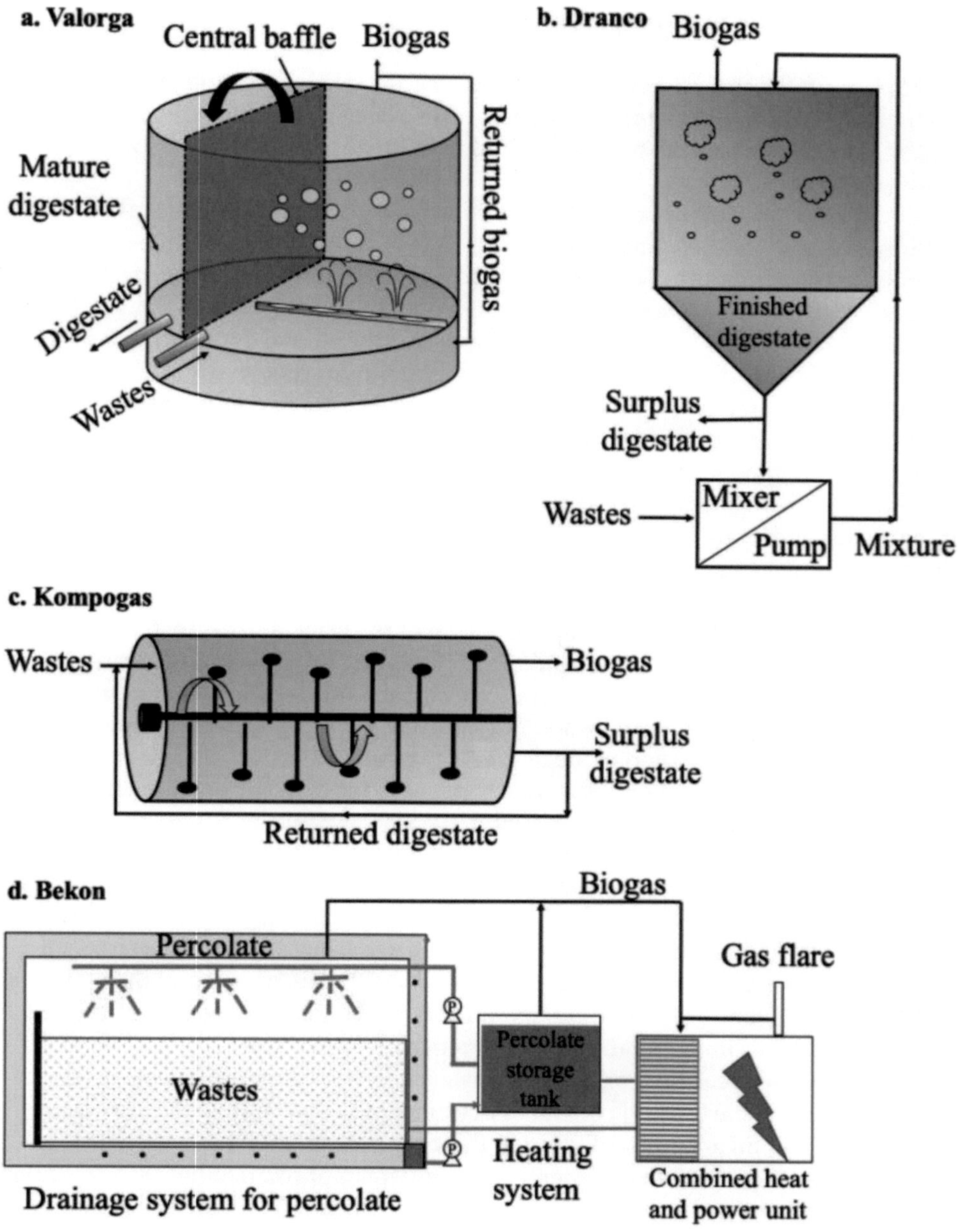

Figure 1.
Dry anaerobic digester designs.

for the leachate collection) and stored at the percolate digester for recycling. Mass transfer in the digester can be promoted by this cycling. Biogas collected from digester and percolate digester is converted into electricity in combined heat and power units (CHP) directly. Digestion period of Bekon process ranges from 28 days to 35 days, and approximately 130 m^3/t of biogas can be recovered [40, 47].

Almost similar to Bekon process, Bioferm (Germany) process also performs the treatment using "garage-type" digester. However, only mesophilic digestion is conducted in Bioferm process, while both mesophilic and thermophilic digestions are conducted in Bekon process. Bioferm process generally operates with a TS content of 25% and a digestion period of 28 days [39].

3.3 New efforts for operating dry anaerobic digestion process

More recently, several new operations of dry anaerobic digestion digesters with some modifications in reactor structure have also been developed, which exhibited high efficiency of methane production and performance stability in dry anaerobic co-digestion.

Zeshan et al. developed a new type of continuous digester, which is called inclined thermophilic dry anaerobic digestion (ITDAD) system [48]. Their pilot-scale experiments indicated that the maximum specific methane yield was 327 L kg^{-1} VS added at total ammonia nitrogen (TAN) of 1895 mg L^{-1} and TS content of 18% (**Table 1**). Kim and Oh proposed a horizontal-type cylindrical continuous digester for the co-digestion of high solids of food waste with paper waste or animal manure [49]. The reactor operates with a TS content of the input wastes ranging from 30 to 50%, and SRT ranges from 30 days to 100 days. 250 L g^{-1} COD_{added} of methane can be recovered when the reactor was applied to co-digestion of food waste with paper waste at SRT of 40 days and 40% of TS content under mesophilic conditions (**Table 1**). The performance they obtained was comparable to the conventional wet digestion and thermophilic dry anaerobic digestion processes.

In terms of liquid recirculation during batch dry anaerobic digestion, most previous studies have focused on optimization of the leachate-to-substrate ratio, the recirculated leachate volume, and recirculation frequency [38, 50]. Meng et al. tested two liquid circulation modes (percolation and immersion) during batch thermophilic dry anaerobic digestion of rice straw using pig urine for liquid circulation [51]. In the percolation mode, leachate was poured on the rice straw-filled mesh bag, while liquid content was passed through the bag. For immersion, the rice straw-filled mesh bag was immersed in the leachate for the designated contact time. Leachate recirculation by percolation might cause nonuniform leachate flow because of the heterogeneous structure of the medium [52], while it is expected that most of the substrate in the bag could be in contact with the leachate by immersion. The methane yield of the immersion mixture of rice straw and solid digestate into leachate was higher than that of percolation of leachate. Furthermore, the methane yield increased from 1 to 24 h of the immersion period, while it decreased after longer than 24 h of immersion. Therefore, pig urine can be used as liquid recirculation medium under certain conditions. However, large-scale validation is needed.

Moreover, the startup and control of dry anaerobic digestion tends to be more difficult than liquid anaerobic digestion, due to the low mass transfer in dry anaerobic digestion [34]. In commercial dry digester, approximately 50–70% of the finished digestate need to be reused as inoculum, which reduces the efficiency of waste treatment [53]. Recently, several studies have pointed that the finished material (effluent) from liquid anaerobic digesters is the best inoculum for dry anaerobic digestion [53, 54]. This is because liquid digestate can provide supplement nitrogen, water, trace elements, and alkalinity to the system [55, 56]. Xu et al. [57] compared the performance of the dry anaerobic digestion yard trimming of using solid digestate and dewatered effluent from liquid anaerobic digester as inoculum. They found that comparable methane yield and volumetric methane productivities are generated at each F/I ratio (0.2–2, based on TS weight) when conducting the digestion using these two kinds of inoculum, while startup time is reduced using dewatered effluent as inoculum. However, the studies are limited in laboratory scale; liquid anaerobic effluent has not been applied in commercial-scale dry anaerobic digestion process, due to the difficult transportation of liquid digestate (effluent) to dry anaerobic digestion plant. A pilot-scale integrated anaerobic digestion process by combining liquid anaerobic digestion and dry anaerobic digestion has been reported in Li et al. [58]. Liquid anaerobic digestion and dry anaerobic digestion are constructed side by side, and liquid digestate is used as inoculum for dry anaerobic digestion. However, larger-scale studies should be considered in the future studies for doing the better choice.

4. Digestate from dry digestion for soil amendment

4.1 Nitrate leaching risk after biogas digestate amendment

Anaerobic digestion is the degradation of organic substrates to biogas and produces a by-product "anaerobic digestate" which contains significant amounts of mineral nitrogen (N), which is available for plants [59]. Biogas digestate typically has a high concentration of ammonium (NH_4^+) and relatively little carbon (C), with NH_4^+-N accounting for 35–81% of total N and a C/N ratio of 2.0–24.8 [15, 18]. Moreover, it contains other macro- and micronutrients that are necessary for plant growth [7, 60].

The merits and demerits in the application of biogas digestate have been addressed in numerous papers. For example, the benefits are to improve the soil properties by reducing the bulk density, to increase the saturated hydraulic conductivity and the moisture retention capacity [61, 62], to sustain soil organic matter concentrations [63, 64], to enhance biological activities [59, 65, 66], and to suppress pathogenic organisms [15, 67]. In contrast, the demerits are to enhance nitrate leaching risk and to bring chemical and biological contaminations, such as heavy metals, organic pollutants [15, 68–70], *Salmonella*, and *Escherichia coli*, which are the most prevalent pathogenic microorganisms found in manures [71, 72].

Once biogas digestate is applied to a field, the NH_4^+-N is subjected to different processes: volatilization, absorption by clay particles, take-up by plants, immobilization into soil organic matter, and/or nitrification [73]. In general, NH_4^+-N in biogas digestate is readily nitrified to nitrate (NO_3^-) in soil [74–76]. Since few NO_3^- can be absorbed by soil particles, most of excess NO_3^- moves downward with drainage water and is eventually leached from the soil profile [77]. Many studies have reported the application of biogas digestate enhances NO_3^- leaching risk in the soil [76, 78–80]. In particular, the nitrate leaching potential is much higher in soil with neutral pH soil than in soil with lower pH [81, 82].

Stumborg [83] reported that dynamics of inorganic N, especially NO_3^-, is directly influenced by the soil type, climate, frequency of application, and property of the digestate. Rigby and Smith [84] conducted a laboratory experiment to investigate the effect of digestate deriving from different waste types (industrial, agricultural, and municipal solid waste or sewage sludge) on the N dynamics in three types of soil (sandy loam, sandy silt loam, and silty clay) and found that NO_3^- concentration was higher in sandy loam and NO_3^- did not accumulate in silty clay soil due to denitrification. Therefore, it is necessary to consider nitrate leaching risk in applying biogas digestate to an agricultural field from different aspects, such as the properties of digestate, soil type, and moisture content.

4.2 Biogas digestate mixed with crop residue to mitigate nitrate leaching risk

Several management strategies have been proposed to mitigate nitrate leaching: (i) limiting N application rates, (ii) synchronizing N supply to plant demand, (iii) adopting cover crop techniques, (iv) using nitrification inhibitors, and (v) applying a C source such as wheat or rice straw [85]. Incorporating digestate with straw residue from harvested crops is a promising practice to retain NO_3^- in the soil. Crop residue with a low C/N ratio degrades fast [86, 87], which increases the soil microbial biomass [88] and stimulates net N mineralization [87, 89]. In contrast, crop residue with a high C/N ratio stimulates net N immobilization, leading to a lower risk of NO_3^- leaching [90]. Previous studies have indicated that applying organic manure [91] or mineral N fertilizer [92] with straw (high C/N ratio) into cultivated soils reduced the accumulation of NO_3^- in the soil, since soil microbes use labile C contained in straw as an energy and carbon source with rapid microbial N immobilization [93, 94], thus decreasing NO_3^- leaching [95].

Wang [82] showed that NO_3^- concentration was the lowest in the treatment of biogas digestate mixed with a high amount of rice straw to adjust the C/N ratio from 12 to 30 (Mix2). The NO_3^- concentration in soil was much lower in Mix2 for a 90-day incubation period than in the other treatments, such as only biogas digestate and chemical fertilizer, indicating that most of the N added to Mix2 was microbially immobilized. Other studies also indicated that application of straw induced net N immobilization during the initial stages and released N at a later stage and the timing is largely dependent on climatic and soil factors including soil fertility [96–98]. It has been reported that application of crop residues reduces N losses and causes greater N retention in soil [99]. Yang [94] showed from a 5-year field experiment that straw application reduced soil NO_3^- leaching losses by 13% compared with the control treatment.

It is a matter of concern when N transformation process changes from immobilization to mineralization. In Kikugawa soil (pH = 7.0), the markedly low NO_3^- in Mix2 started to increase from day 35, indicating the net re-mineralization of the once immobilized N and soil organic N from day 35. In contrast, in Fuchu soil (pH = 5.7), NO_3^- started to increase only after day 60, indicating that microbial immobilization consistently dominated the nitrogen cycling process for the first 60 days. The period of N retention and N supply processes differ among soils [100]. Zhao [101] reported that N retention was much longer in a soil with lower pH (5.3) than in a soil with neutral pH (7.6). Soil fertility may also be involved in the change from N immobilization to N mineralization, since Pan [95] reported that N mineralization starts earlier in a fertile soil after the occurrence of N immobilization. Kikugawa soil (total C: 73.2 g kg^{-1} soil) showed higher fertility than Fuchu soil (total C: 35 g C kg^{-1} soil), and thus the earlier change from N immobilization to N mineralization occurred in fertile Kikugawa soil.

4.3 Effect of biogas digestate application on root-knot nematode

Root-knot nematodes (*Meloidogyne* spp.) are the most economically damaging group of plant-parasitic nematodes (PPNs) worldwide [102–104]. The genus *Meloidogyne* is composed of approximately 100 species and parasitizes thousands of plant species [105, 106]. This parasitism results in poor host plant growth and presents a serious threat to the production of many important horticultural and field crops [107–109]. As countermeasures, several means with nematode-suppressive properties have been reported, such as applications of compost with a low C/N ratio (< 20) [110, 111], volatile fatty acids [112], chitin [113], and plant-specific toxins [114]. A few studies also showed that application of biogas digestate to soil reduced the root gall formation of root-knot nematodes of tomato [115] and the damage to sugar beet by *Heterodera schachtii* [116].

A recent study showed that populations of *M. incognita* did not decrease in soil added with dry biogas digestate (C/N ratio of 20) treatment, compared with those in chemical fertilizer treatment [82]. Several studies have already reported that not all types of organic amendments are beneficial in the suppression of root-knot nematodes [117, 118]. For instance, Bulluck [117] also observed that

M. incognita populations were not affected by amendments of swine manure and composts. There are several factors which determine the effect of organic fertilizer on plant-parasitic nematodes, and the most commonly reported one is C/N ratio [119]. Organic amendment with a C/N ratio in the range of 15–20 was considered most effective [114]. In a study by Agu [120], plants of African yam bean treated with poultry and farmyard manures (C/N ratio of 4 to 12) showed a lower degree of disease caused by root-knot nematodes than those with other organic manures with C/N ratios higher than 30. In the study by Wang [82], the populations of *M. incognita* drastically decreased in Mix 2 treatment, in which biogas digestate was co-added with rice straw to increase its C/N ratio from 12 to 30.

Organic amendment may have different effects on different soil microbial groups, and nematodes could be reduced by such a modified microbial group [119, 121]. The prokaryotic community structure of the treatments reported by Wang [82] was evaluated, and the results showed that Mix2 treatment, in which low NO_3^- risk and high nematode suppression were confirmed, was separated from the other treatments, indicating that a specific microbial community was developed in the treatment (**Figure 2**). Several papers have already reported that the application of biogas digestate affected the community structure of bacteria and fungi [122–124]. In general, organic amendment stimulates a broad range of (micro) organisms involved in the soil food web, many of which are potential predators, such as diplogasterid [125] and dorylaimid [126], or invertebrate antagonists, such as enchytraeids and earthworms [127]. Moreover, nematode suppression might result from increased incidences and levels of nematode-antagonistic fungi following amendment application. According to Wang [128, 129], the application of sunn hemp crop residues to soil decreased the population levels of the plant-parasitic nematode *Rotylenchulus reniformis* and increased levels of nematode-trapping fungi, such as *Arthrobotrys oligospora* [130] and *Ematoctonus leiosporus* [131]. The mode of action in biogas digestate leading to nematode suppression and stimulation of microorganisms is complex and dependent on the nature of the original wastes. Therefore, long-term use of biogas digestate to build suppressive elements of the soil food web remains an elusive goal.

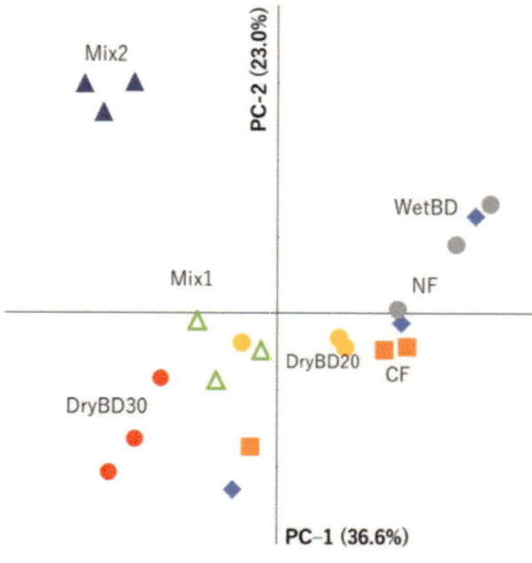

	NF	CF	WetBD	DryBD20	DryBD30	Mix1	Mix2
Alphaproteobacteria	10.1	10.7	10.3	10.5	10.5	11.1	11.1
Bacilli	8.2	8.4	8.7	9.3	12.6	8.3	11.5
Actinobacteria	6.1	6.7	6.7	6.8	6.7	6.9	9.4
Betaproteobacteria	7.4	7.2	6.5	6.5	6.4	6.8	5.9
Thermoleophilia	6.8	7.0	6.3	6.3	6.8	6.6	6.0
Deltaproteobacteria	5.4	5.1	5.2	5.0	5.0	5.2	6.0
Planctomycetia	5.3	4.8	5.3	4.9	4.3	4.7	4.8
Gammaproteobacteria	4.0	4.3	4.2	4.2	3.9	4.3	4.6
Thaumarchaeota	3.9	4.5	4.4	4.9	3.9	4.1	3.7
Acidobacteria-6	2.9	3.0	2.9	2.8	2.7	3.0	2.7
Chloroflexi;c__TK10	2.7	2.7	1.3	2.7	3.2	2.8	2.5
Chloroflexi;c__Ellin6529	2.7	2.6	2.6	2.6	2.5	2.6	1.8
Chloroflexi;c__Gitt-GS-136	2.2	2.5	2.2	2.3	2.5	2.3	1.5
Chloroflexi;c__Anaerolineae	2.6	2.1	2.5	2.1	1.8	2.0	1.7
Gemmatimonadetes	2.3	2.3	1.6	2.0	2.2	2.3	1.9
Acidimicrobiia	2.0	1.9	2.1	2.1	2.0	2.1	1.9
Phycisphaerae	2.3	2.0	2.3	1.8	1.4	1.6	1.4
Gemmatimonadetes	1.6	1.8	1.7	1.7	1.6	1.6	1.4
Saprospirae	1.7	1.7	1.8	1.6	1.1	1.4	1.4
Others	17.9	16.8	18.7	17.8	17.1	18.3	16.8
Unassigned	2.1	2.0	2.5	2.3	2.0	2.2	2.0

Figure 2.
A Uni-Frac weighted PCA analysis of prokaryotic communities of soils with different amendments and incubated for 90 days. NF: no fertilizer, CF: chemical fertilizers, DryBD20: dry biogas digestate with an C/N ratio of 12, DryBD30: dry biogas digestate with an C/N ratio of 16, Mix1: DryBD20 mixed with a low amount of rice straw to adjust its C/N ratio to 16, Mix2: DryBD20 with a high amount of rice straw to adjust its C/N ratio to 30.

5. Conclusion

Dry anaerobic digestion is appropriate for treatment of agricultural waste with high TS content. Optimization of C/N ratio, F/I ratio (or organic loading rate), and TS content is key to avoid failure of digestion. Several batch and continuous dry digestion technologies have been already applied in practice, while new techniques have been also proposed. Solid digestate is beneficial to supply nutrient into the soil and improve soil properties. On the other hand, nitrate leaching is one of the concerns of the digestate application. Digestate C/N ratio adjustment by mixing with crop residue can mitigate nitrate leaching. In addition, it can also mitigate root-knot nematode. More study is needed to optimize dry anaerobic digestion and digestate use for sustainable agricultural waste management.

Acknowledgements

The study was supported by TUAT Collaboration Award 2019 and the Environment Research and Technology Development Fund (grant number: 1B-1103 and 1-1404) of the Ministry of the Environment, Japan.

Conflict of interest

The authors declare no conflict of interest.

Author details

Shohei Riya[1*], Lingyu Meng[1], Yuexi Wang[2], Chol Gyu Lee[2], Sheng Zhou[3], Koki Toyota[2] and Masaaki Hosomi[1]

1 Graduate School of Engineering, Tokyo University of Agriculture and Technology, Tokyo, Japan

2 Graduate School of Bio-Applications and Systems Engineering, Tokyo University of Agriculture and Technology, Tokyo, Japan

3 Eco-Environmental Protection Research Institute, Shanghai Academy of Agricultural Sciences, Shanghai, China

*Address all correspondence to: sriya@cc.tuat.ac.jp

References

[1] Mussoline W, Esposito G, Giordano A, Lens P. The anaerobic digestion of rice straw: A review. Critical Reviews in Environmental Science and Technology. 2013;**43**(9):895-915

[2] Kumar R, Singh S, Singh OV. Bioconversion of lignocellulosic biomass: Biochemical and molecular perspectives. Journal of Industrial Microbiology & Biotechnology. 2008;**35**(5):377-391

[3] Potter P, Ramankutty N, Bennett EM, Donner SD. Characterizing the spatial patterns of global fertilizer application and manure production. Earth Interactions. 2010;**14**:1-22

[4] Schievano A, D'Imporzano G, Salati S, Adani F. On-field study of anaerobic digestion full-scale plants (part I): An on-field methodology to determine mass, carbon and nutrients balance. Bioresource Technology. 2011;**102**(17):7737-7744

[5] Elsayed M, Abomohra A, Ai P, Wang DL, El-Mashad HM, Zhang YL. Biorefining of rice straw by sequential fermentation and anaerobic digestion for bioethanol and/or biomethane production: Comparison of structural properties and energy output. Bioresource Technology. 2018;**268**:183-189

[6] Sawatdeenarunat C, Nguyen D, Surendra C, Shrestha S, Rajendran K, Oechsner H, et al. Anaerobic biorefinery: Current status, challenges, and opportunities. Bioresource Technology. 2016;**215**:304-313

[7] Alburquerque JA, de la Fuente C, Ferrer-Costa A, Carrasco L, Cegarra J, Abad M, et al. Assessment of the fertiliser potential of digestates from farm and agroindustrial residues. Biomass and Bioenergy. 2012;**40**:181-189

[8] Lal R. Carbon emission from farm operations. Environment International. 2004;**30**(7):981-990

[9] Ge X, Xu F, Li Y. Solid-state anaerobic digestion of lignocellulosic biomass: Recent progress and perspectives. Bioresource Technology. 2016;**205**:239-249

[10] Li Y, Park SY, Zhu J. Solid-state anaerobic digestion for methane production from organic waste. Renewable and Sustainable Energy Reviews. 2011;**15**(1):821-826

[11] Logan M, Visvanathan C. Management strategies for anaerobic digestate of organic fraction of municipal solid waste: Current status and future prospects. Waste Management & Research. 2019;**37**:27-39

[12] Angelonidi E, Smith SR. A comparison of wet and dry anaerobic digestion processes for the treatment of municipal solid waste and food waste. Water Environment Journal. 2015;**29**(4):549-557

[13] Chiumenti A, da Borso F, Limina S. Dry anaerobic digestion of cow manure and agricultural products in a full-scale plant: Efficiency and comparison with wet fermentation. Waste Management. 2018;**71**:704-710

[14] Moller HB, Sommer SG, Ahring BK. Separation efficiency and particle size distribution in relation to manure type and storage conditions. Bioresource Technology. 2002;**85**(2):189-196

[15] Nkoa R. Agricultural benefits and environmental risks of soil fertilization with anaerobic digestates: A review. Agronomy for Sustainable Development. 2014;**34**(2):473-492

[16] Riya S, Suzuki K, Meng L, Zhou S, Terada A, Hosomi M. The influence of the total solid content on the stability of dry-thermophilic anaerobic digestion of rice straw and pig manure. Waste Management. 2018;**76**:350-356

[17] Zhou S, Nikolausz M, Zhang JN, Riya SH, Terada A, Hosomi M. Variation of the microbial community in thermophilic anaerobic digestion of pig manure mixed with different ratios of rice straw. Journal of Bioscience and Bioengineering. 2016;**122**(3):334-340

[18] Möller K, Müller T. Effects of anaerobic digestion on digestate nutrient availability and crop growth: A review. Engineering in Life Sciences. 2012;**12**(3):242-257

[19] Rajagopal R, Masse DI, Singh G. A critical review on inhibition of anaerobic digestion process by excess ammonia. Bioresource Technology. 2013;**143**:632-641

[20] Fernandes TV, Keesman KJ, Zeeman G, van Lier JB. Effect of ammonia on the anaerobic hydrolysis of cellulose and tributyrin. Biomass and Bioenergy. 2012;**47**:316-323

[21] Kayhanian M. Ammonia inhibition in high-solids biogasification: An overview and practical solutions. Environmental Technology. 1999;**20**(4):355-365

[22] Li Y, Zhu J, Wan C, Park SY. Solid-state anaerobic digestion of corn stover for biogas production. Transactions of the Asabe. 2011;**54**(4):1415-1421

[23] Abouelenien F, Namba Y, Kosseva MR, Nishio N, Nakashimada Y. Enhancement of methane production from co-digestion of chicken manure with agricultural wastes. Bioresource Technology. 2014;**159**:80-87

[24] Abouelenien F, Kitamura Y, Nishio N, Nakashimada Y. Dry anaerobic ammonia-methane production from chicken manure. Applied Microbiology and Biotechnology. 2009;**82**(4):757-764

[25] Li YQ, Zhang RH, Chen C, Liu GQ, He YF, Liu XY. Biogas production from co-digestion of corn Stover and chicken manure under anaerobic wet, hemi-solid, and solid state conditions. Bioresource Technology. 2013;**149**:406-412

[26] Zhou Y, Li C, Nges IA, Liu J. The effects of pre-aeration and inoculation on solid-state anaerobic digestion of rice straw. Bioresource Technology. 2017;**224**:78-86

[27] Meng LY, Xie L, Kinh CT, Suenaga T, Hori T, Riya S, et al. Influence of feedstock-to-inoculum ratio on performance and microbial community succession during solid-state thermophilic anaerobic co-digestion of pig urine and rice straw. Bioresource Technology. 2018;**252**:127-133

[28] Li YY, Wang YQ, Yu ZH, Lu JX, Li DY, Wang GY, et al. Effect of inoculum and substrate/inoculum ratio on the performance and methanogenic archaeal community structure in solid state anaerobic co-digestion of tomato residues with dairy manure and corn Stover. Waste Management. 2018;**81**:117-127

[29] Ma XG, Jiang T, Chang JL, Tang Q, Luo T, Cui ZJ. Effect of substrate to inoculum ratio on biogas production and microbial community during hemi-solid-state batch anaerobic Co-digestion of rape straw and dairy manure. Applied Biochemistry and Biotechnology. 2019;**189**(3):884-902

[30] Cui ZF, Shi J, Li YB. Solid-state anaerobic digestion of spent wheat straw from horse stall. Bioresource Technology. 2011;**102**(20):9432-9437

[31] Suksong W, Kongjan P, Prasertsan P, Imai T, O-Thong S. Optimization and microbial community analysis for production of biogas from solid waste residues of palm oil mill industry by solid-state anaerobic digestion. Bioresource Technology. 2016;**214**:166-174

[32] Chen Y, Wen Y, Zhou JW, Xu C, Zhou Q. Effects of pH on the hydrolysis of lignocellulosic wastes and volatile fatty acids accumulation: The contribution of biotic and abiotic factors. Bioresource Technology. 2012;**110**:321-329

[33] Xu FQ, Wang ZW, Tang L, Li YB. A mass diffusion-based interpretation of the effect of total solids content on solid-state anaerobic digestion of cellulosic biomass. Bioresource Technology. 2014;**167**:178-185

[34] Le Hyaric R, Benbelkacem H, Bollon J, Bayard R, Escudie R, Buffiere P. Influence of moisture content on the specific methanogenic activity of dry mesophilic municipal solid waste digestate. Journal of Chemical Technology and Biotechnology. 2012;**87**(7):1032-1035

[35] Bollon J, Benbelkacem H, Gourdon R, Buffiere P. Measurement of diffusion coefficients in dry anaerobic digestion media. Chemical Engineering Science. 2013;**89**:115-119

[36] Abbassi-Guendouz A, Brockmann D, Trably E, Dumas C, Delgenes J-P, Steyer J-P, et al. Total solids content drives high solid anaerobic digestion via mass transfer limitation. Bioresource Technology. 2012;**111**:55-61

[37] Zhu J, Yang L, Li Y. Comparison of premixing methods for solid-state anaerobic digestion of corn Stover. Bioresource Technology. 2015;**175**:430-435

[38] Degueurce A, Tomas N, Le Roux S, Martinez J, Peu P. Biotic and abiotic roles of leachate recirculation in batch mode solid-state anaerobic digestion of cattle manure. Bioresource Technology. 2016;**200**:388-395

[39] Andre L, Pauss A, Ribeiro T. Solid anaerobic digestion: State-of-art, scientific and technological hurdles. Bioresource Technology. 2018;**247**: 1027-1037

[40] Lei ZF, Zhang ZY, Huang WW, Cai W. Recent Progress on dry anaerobic digestion of organic solid wastes: Achievements and challenges. Current Organic Chemistry. 2015;**19**(5):400-412

[41] Rapport J, Zhang R, Jenkins B, Williams R. Current Anaerobic Digestion Technologies Used for Treatment of Municipal Organic Solid Waste. Davis: University of California; Contractor Report to the California Integrated Waste Management Board; 2008

[42] Valorga, References I. Available from: http://www.valorgainternational.fr/en/pag8-OUR-REFERENCES.html

[43] de Laclos HF, Desbois S, Saint-Joly C. Anaerobic digestion of municipal solid organic waste: Valorga full-scale plant in Tilburg, the Netherlands. Water Science and Technology. 1997;**36**(6-7):457-462

[44] Angelidaki I, Ellegaard L, Ahring B. Applications of the Anaerobic Digestion Process. Berlin, Heidelberg: Springer; 2003

[45] Nichols C. Overview of anaerobic digestion technologies in Europe. BioCycle. 2004;**45**(1):47-53

[46] Fu YR, Luo T, Mei ZL, Li J, Qiu K, Ge YH. Dry anaerobic digestion technologies for agricultural straw and acceptability in China. Sustainability. 2018;**10**(12):4588

[47] Bekon. Generating Energy from Biomass. Available from: https://www.bekon.eu/en/generating-energy-from-biomass/

[48] Zeshan, Karthikeyan OP, Visvanathan C. Effect of C/N ratio and ammonia-N accumulation in a

pilot-scale thermophilic dry anaerobic digester. Bioresource Technology. 2012;**113**:294-302

[49] Kim DH, Oh SE. Continuous high-solids anaerobic co-digestion of organic solid wastes under mesophilic conditions. Waste Management. 2011;**31**(9-10):1943-1948

[50] Benbelkacem H, Bayard R, Abdelhay A, Zhang Y, Gourdon R. Effect of leachate injection modes on municipal solid waste degradation in anaerobic bioreactor. Bioresource Technology. 2010;**101**(14):5206-5212

[51] Meng L, Maruo K, Xie L, Riya S, Terada A, Hosomi M. Comparison of leachate percolation and immersion using different inoculation strategies in thermophilic solid-state anaerobic digestion of pig urine and rice straw. Bioresource Technology. 2019;**277**:216-220

[52] Andre L, Durante M, Pauss A, Lespinard O, Ribeiro T, Lamy E. Quantifying physical structure changes and non-uniform water flow in cattle manure during dry anaerobic digestion process at lab scale: Implication for biogas production. Bioresource Technology. 2015;**192**:660-669

[53] Karthikeyan OP, Visvanathan C. Bio-energy recovery from high-solid organic substrates by dry anaerobic bio-conversion processes: A review. Reviews in Environmental Science and Bio-Technology. 2013;**12**(3):257-284

[54] Forster-Carneiro T, Perez M, Romero LI, Sales D. Dry-thermophilic anaerobic digestion of organic fraction of the municipal solid waste: Focusing on the inoculum sources. Bioresource Technology. 2007;**98**(17):3195-3203

[55] Yang L, Xu F, Ge X, Li Y. Challenges and strategies for solid-state anaerobic digestion of lignocellulosic biomass. Renewable and Sustainable Energy Reviews. 2015;**44**:824-834

[56] Shi J, Xu FQ, Wang ZJ, Stiverson JA, Yu ZT, Li YB. Effects of microbial and non-microbial factors of liquid anaerobic digestion effluent as inoculum on solid-state anaerobic digestion of corn Stover. Bioresource Technology. 2014;**157**:188-196

[57] Xu FQ, Wang F, Lin L, Li YB. Comparison of digestate from solid anaerobic digesters and dewatered effluent from liquid anaerobic digesters as inocula for solid state anaerobic digestion of yard trimmings. Bioresource Technology. 2016;**200**:753-760

[58] Li Y, Zhu J, Wan C, Combined liquid to solid-phase anaerobic digestion for methane production from municipal and agricultural wastes. United States. 2014. Patent No. US 8,771,980 B2

[59] Makádi M, Tomócsik A, Orosz V. Digestate: A New Nutrient Source-Review. IntechOpen: Biogas; 2012

[60] Alburquerque JA, de la Fuente C, Bernal MP. Chemical properties of anaerobic digestates affecting C and N dynamics in amended soils. Agriculture, Ecosystems and Environment. 2012;**160**:15-22

[61] Beni C, Servadio P, Marconi S, Neri U, Aromolo R, Diana G. Anaerobic digestate administration: Effect on soil physical and mechanical behavior. Communications in Soil Science and Plant Analysis. 2012;**43**(5):821-834

[62] Garg RN, Pathak H, Das D, Tomar R. Use of flyash and biogas slurry for improving wheat yield and physical properties of soil. Environmental Monitoring and Assessment. 2005;**107**(1-3):1-9

[63] Arthurson V. Closing the global energy and nutrient cycles through application of biogas residue to agricultural land–potential benefits and drawback. Energies. 2009;**2**(2):226-242

[64] Sreekrishnan T, Kohli S, Rana V. Enhancement of biogas production from solid substrates using different techniques–A review. Bioresource Technology. 2004;**95**(1):1-10

[65] Scherr KE, Lundaa T, Klose V, Bochmann G, Loibner AP. Changes in bacterial communities from anaerobic digesters during petroleum hydrocarbon degradation. Journal of Biotechnology. 2012;**157**(4):564-572

[66] Ziemiński K, Frąc M. Methane fermentation process as anaerobic digestion of biomass: Transformations, stages and microorganisms. African Journal of Biotechnology. 2012;**11**(18):4127-4139

[67] Insam H, Gómez-Brandón M, Ascher J. Manure-based biogas fermentation residues–friend or foe of soil fertility? Soil Biology and Biochemistry. 2015;**84**:1-14

[68] Cheng J, Chen Y, He T, Liao R, Liu R, Yi M, et al. Soil nitrogen leaching decreases as biogas slurry DOC/N ratio increases. Applied Soil Ecology. 2017;**111**:105-113

[69] Sänger A, Geisseler D, Ludwig B. Effects of moisture and temperature on greenhouse gas emissions and C and N leaching losses in soil treated with biogas slurry. Biology and Fertility of Soils. 2011;**47**(3):249-259

[70] Svoboda N, Taube F, Wienforth B, Kluß C, Kage H, Herrmann A. Nitrogen leaching losses after biogas residue application to maize. Soil and Tillage Research. 2013;**130**:69-80

[71] Bicudo JR, Goyal SM. Pathogens and manure management systems: A review. Environmental Technology. 2003;**24**(1):115-130

[72] Hutchison M, Walters LD, Avery S, Munro F, Moore A. Analyses of livestock production, waste storage, and pathogen levels and prevalences in farm manures. Applied and Environmental Microbiology. 2005;**71**(3):1231-1236

[73] Amlinger F, Götz B, Dreher P, Geszti J, Weissteiner C. Nitrogen in biowaste and yard waste compost: Dynamics of mobilisation and availability—A review. European Journal of Soil Biology. 2003;**39**(3):107-116

[74] Johansen A, Carter MS, Jensen ES, Hauggard-Nielsen H, Ambus P. Effects of digestate from anaerobically digested cattle slurry and plant materials on soil microbial community and emission of CO_2 and N_2O. Applied Soil Ecology. 2013;**63**:36-44

[75] Sänger A, Geisseler D, Ludwig B. C and N dynamics of a range of biogas slurries as a function of application rate and soil texture: A laboratory experiment. Archives of Agronomy and Soil Science. 2014;**60**(12):1779-1794

[76] Sawada K, Toyota K. Effects of the application of Digestates from wet and dry anaerobic fermentation to Japanese Paddy and Upland soils on short-term nitrification. Microbes and Environments. 2015;**30**(1):37-43

[77] Goberna M, Podmirseg SM, Waldhuber S, Knapp BA, García C, Insam H. Pathogenic bacteria and mineral N in soils following the land spreading of biogas digestates and fresh manure. Applied Soil Ecology. 2011;**49**:18-25

[78] Cavalli D, Corti M, Baronchelli D, Bechini L, Marino Gallina P. CO_2 emissions and mineral nitrogen dynamics following application to soil of undigested liquid cattle manure and digestates. Geoderma. 2017;**308**:26-35

[79] Gómez-Brandón M, Juárez MF-D, Zangerle M, Insam H. Effects of digestate on soil chemical and

microbiological properties: A comparative study with compost and vermicompost. Journal of Hazardous Materials. 2016;**302**:267-274

[80] Viaene J, Agneessens L, Capito C, Ameloot N, Reubens B, Willekens K, et al. Co-ensiling, co-composting and anaerobic co-digestion of vegetable crop residues: Product stability and effect on soil carbon and nitrogen dynamics. Scientia Horticulturae. 2017;**220**:214-225

[81] Herrmann A, Kage H, Taube F, Sieling K. Effect of biogas digestate, animal manure and mineral fertilizer application on nitrogen flows in biogas feedstock production. European Journal of Agronomy. 2017;**91**:63-73

[82] Wang Y, Chikamatsu S, Gegen T, Sawada K, Toyota K, Riya S, et al. Application of biogas digestate with rice straw mitigates nitrate leaching potential and suppresses root-knot nematode (*Meloidogyne incognita*). Agronomy. 2019;**9**(5):227

[83] Stumborg C, Schoenau JJ, Malhi SS. Nitrogen balance and accumulation pattern in three contrasting prairie soils receiving repeated applications of liquid swine and solid cattle manure. Nutrient Cycling in Agroecosystems. 2007;**78**(1):15-25

[84] Rigby H, Smith SR. Nitrogen availability and indirect measurements of greenhouse gas emissions from aerobic and anaerobic biowaste digestates applied to agricultural soils. Waste Management. 2013;**33**(12): 2641-2652

[85] Di HJ, Cameron KC. Nitrate leaching in temperate agroecosystems: Sources, factors and mitigating strategies. Nutrient Cycling in Agroecosystems. 2002;**64**(3):237-256

[86] Pansu M, Thuriès L. Kinetics of C and N mineralization, N immobilization and N volatilization of organic inputs in soil. Soil Biology and Biochemistry. 2003;**35**(1):37-48

[87] Yanni SF, Whalen JK, Simpson MJ, Janzen HH. Plant lignin and nitrogen contents control carbon dioxide production and nitrogen mineralization in soils incubated with Bt and non-Bt corn residues. Soil Biology and Biochemistry. 2011;**43**(1):63-69

[88] Hoyle FC, Murphy DV. Influence of organic residues and soil incorporation on temporal measures of microbial biomass and plant available nitrogen. Plant and Soil. 2011;**347**(1-2):53

[89] Puttaso A, Vityakon P, Saenjan P, Trelo-Ges V, Cadisch G. Relationship between residue quality, decomposition patterns, and soil organic matter accumulation in a tropical sandy soil after 13 years. Nutrient Cycling in Agroecosystems. 2011;**89**(2):159-174

[90] Trinsoutrot I, Recous S, Bentz B, Lineres M, Cheneby D, Nicolardot B. Biochemical quality of crop residues and carbon and nitrogen mineralization kinetics under nonlimiting nitrogen conditions. Soil Science Society of America Journal. 2000;**3**(64):918-926. Available from: https://doi.org/10.2136/sssaj2000.643918x

[91] Demiraj E, Libutti A, Malltezi J, Rroço E, Brahushi F, Monteleone M, et al. Effect of organic amendments on nitrate leaching mitigation in a sandy loam soil of Shkodra district, Albania. Italian Journal of Agronomy. 2018;**13**(1):93-102

[92] Gai X, Liu H, Liu J, Zhai L, Wang H, Yang B, et al. Contrasting impacts of long-term application of manure and crop straw on residual nitrate-N along the soil profile in the North China plain. Science of The Total Environment. 2019;**650**:2251-2259

[93] Shindo H, Nishio T. Immobilization and remineralization

of N following addition of wheat straw into soil: Determination of gross N transformation rates by 15N-ammonium isotope dilution technique. Soil Biology and Biochemistry. 2005;**37**(3):425-432

[94] Yang S, Wang Y, Liu R, Xing L, Yang Z. Improved crop yield and reduced nitrate nitrogen leaching with straw return in a rice-wheat rotation of Ningxia irrigation district. Scientific Reports. 2018;**8**(1):9458

[95] Pan F-F, Yu W-T, Ma Q, Zhou H, Jiang C-M, Xu Y-G, et al. Influence of 15 N-labeled ammonium sulfate and straw on nitrogen retention and supply in different fertility soils. Biology and Fertility of Soils. 2017;**53**(3):303-313

[96] Aulakh M, Walters D, Doran J, Francis D, Mosier A. Crop residue type and placement effects on denitrification and mineralization. Soil Science Society of America Journal. 1991;**55**(4):1020-1025

[97] Chen B, Liu E, Tian Q, Yan C, Zhang Y. Soil nitrogen dynamics and crop residues. A review. Agronomy for Sustainable Development. 2014;**34**(2):429-442

[98] Nicolardot B, Chaussod R, Cheneby D, Baugnet M. Effects of some factors on immobilization and remineralization of nitrogen in soils. In: Stable Isotopes in Plant Nutrition, Soil Fertility and Environmental Studies. Proceedings of an International Symposium on the Use of Stable Isotopes in Plant Nutrition, Soil Fertility and Environmental Studies. Vienna: IAEA, FAO; 1991. pp. 327-330. https://inis.iaea.org/search/search.aspx?orig_q=RN:220889251991

[99] Delgado J. Quantifying the loss mechanisms of nitrogen. Journal of Soil and Water Conservation. 2002;**57**(6):389-398

[100] Yadvinder S, Gupta RK, Gurpreet S, Jagmohan S, Sidhu HS, Bijay S. Nitrogen and residue management effects on agronomic productivity and nitrogen use efficiency in rice–wheat system in Indian Punjab. Nutrient Cycling in Agroecosystems. 2009;**84**(2):141-154

[101] Zhao Y, Zhang J, Müller C, Cai Z. Temporal variations of crop residue effects on soil N transformation depend on soil properties as well as residue qualities. Biology and Fertility of Soils. 2018;**54**(5):659-669

[102] Kayani MZ, Mukhtar T, Hussain MA. Effects of southern root knot nematode population densities and plant age on growth and yield parameters of cucumber. Crop Protection. 2017;**92**:207-212

[103] Ralmi NHAA, Khandaker MM, Mat N. Occurrence and control of root knot nematode in crops: A review. Australian Journal of Crop Science. 2016;**11**(12):1649

[104] Yadav U. Recent trends in nematode management practices: The Indian context. International Research Journal of Engineering and Technology. 2017;**12**:482-489

[105] Eisenback JD, Triantaphyllou HH. Root-knot nematodes: Meloidogyne species and races. Manual of Agricultural Nematology. 1991;**1**:191-274

[106] Hunt DJ, Handoo ZA. Taxonomy, identification and principal species. Root-knot Nematodes. 2009;**1**:55-88

[107] Fujimoto T, Hasegawa S, Otobe K, Mizukubo T. Effect of water flow on the mobility of the root-knot nematode *Meloidogyne incognita* in columns filled with glass beads, sand or andisol. Applied Soil Ecology. 2009;**43**(2-3):200-205

[108] Seo Y, Kim YH. Control of *Meloidogyne incognita* using mixtures

of organic acids. The Plant Pathology Journal. 2014;**30**(4):450

[109] Shah SJ, Anjam MS, Mendy B, Anwer MA, Habash SS, Lozano-Torres JL, et al. Damage-associated responses of the host contribute to defence against cyst nematodes but not root-knot nematodes. Journal of Experimental Botany. 2017;**68**(21-22):5949-5960

[110] Castagnone-Sereno P, Kermarrec A. Invasion of tomato roots and reproduction of *Meloidogyne incognita* as affected by raw sewage sludge. Journal of Nematology. 1991;**23**(4S):724

[111] Nico AI, Jiménez-Díaz RM, Castillo P. Control of root-knot nematodes by composted agro-industrial wastes in potting mixtures. Crop Protection. 2004;**23**(7):581-587

[112] Mahran A, Conn K, Tenuta M, Lazarovits G, Daayf F. Effectiveness of liquid hog manure and acidification to kill Pratylenchus spp. in soil. Journal of Nematology. 2008;**40**(4):266

[113] Hallmann J, Rodríguez-Kábana R, Kloepper JW. Chitin-mediated changes in bacterial communities of the soil, rhizosphere and within roots of cotton in relation to nematode control. Soil Biology and Biochemistry. 1999;**31**(4):551-560

[114] Mcsorley R. Overview of organic amendments for management of plant-parasitic nematodes, with case studies from Florida. Journal of Nematology. 2011;**43**(2):69

[115] Jothi G, Pugalendhi S, Poornima K, Rajendran G. Management of root-knot nematode in tomato *Lycopersicon esculentum*, mill., with biogas slurry. Bioresource Technology. 2003;**89**(2):169-170

[116] Westphal A, Kücke M, Heuer H. Soil amendment with digestate from bio-energy fermenters for mitigating damage to *Beta vulgaris* subspp. By *Heterodera schachtii*. Applied Soil Ecology. 2016;**99**:129-136

[117] Bulluck Iii L, Barker K, Ristaino J. Influences of organic and synthetic soil fertility amendments on nematode trophic groups and community dynamics under tomatoes. Applied Soil Ecology. 2002;**21**(3):233-250

[118] Korayem A. Effect of some organic wastes on *Meloidogyne incognita* development and tomato tolerance to the nematode. Egyptian Journal of Phytopathology. 2003;**31**(1-2):119-127

[119] Renčo M. Organic amendments of soil as useful tools of plant parasitic nematodes control. Helminthologia. 2013;**50**(1):3-14

[120] Agu C. Effects of organic manure types on root-gall nematode disease and African yam bean yield. Journal of American Science. 2008;**4**(1):76-79

[121] Akhtar M, Malik A. Roles of organic soil amendments and soil organisms in the biological control of plant-parasitic nematodes: A review. Bioresource Technology. 2000;**74**(1):35-47

[122] Garcia-Sánchez M, Garcia-Romera I, Cajthaml T, Tlustoš P, Száková J. Changes in soil microbial community functionality and structure in a metal-polluted site: The effect of digestate and fly ash applications. Journal of Environmental Management. 2015;**162**:63-73

[123] Gielnik A, Pechaud Y, Huguenot D, Cébron A, Riom J-M, Guibaud G, et al. Effect of digestate application on microbial respiration and bacterial communities' diversity during bioremediation of weathered petroleum hydrocarbons contaminated soils. Science of The Total Environment. 2019;**670**:271-281

[124] Hupfauf S, Bachmann S, Juárez MF-D, Insam H, Eichler-Löbermann B. Biogas digestates affect crop P uptake and soil microbial community composition. The Science of the Total Environment. 2016;**542**:1144-1154

[125] Yeates G. Predation by Mononchoides potohikus (Nematoda: Diplogasteridae) in laboratory culture. Nematologica. 1969;**15**(1):1-9

[126] Esser R. Biological Control of Nematodes by Nematodes: Dorylaims (Nematoda: Dorylaimina). I. Gainesville: Fla. Department Agric. & Consumer Serv., Division of Plant Industry; 1987

[127] Yeates G. Soil nematode populations depressed in the presence of earthworms. Pedobiologia. 1981;**22**(1):191-195

[128] Wang K-H, Sipes B, Schmitt D. Suppression of *Rotylenchulus reniformis* by *Crotalaria juncea*, *Brassica napus*, and *Tagetes erecta*. Nematropica. 2001;**31**(2):235-249

[129] Wang K-H, Sipes B, Schmitt D. Management of *Rotylenchulus reniformis* in pineapple, *Ananas comosus*, by intercycle cover crops. Journal of Nematology. 2002;**34**(2):106

[130] Bordallo J, Lopez-Llorca L, Jansson HB, Salinas J, Persmark L, Asensio L. Colonization of plant roots by egg-parasitic and nematode-trapping fungi. The New Phytologist. 2002;**154**(2):491-499

[131] Jaffee B, Ferris H, Scow K. Nematode-trapping fungi in organic and conventional cropping systems. Phytopathology. 1998;**88**(4):344-350

Chapter 3

Biogas Production from Water Hyacinth

Winardi Dwi Nugraha, Syafrudin and Lathifah Laksmi Pradita

Abstract

The current existence of water hyacinth as a waterweed is very unsettling and detrimental, so various alternatives were made to utilize its existence. One of the alternatives is biogas fuel. Water hyacinth leaves can be used as biogas fuel because of its cellulose, nitrogen, essential nutrients, and high fermentation contents. Through this chapter, two kinds of methods used to test the optimization of biogas production from water hyacinth leaves will be explained, namely, the liquid anaerobic digestion (L-AD) and solid-state anaerobic digestion (SS-AD) methods using total solid (TS), food to microorganism (F/M), and carbon to nitrogen (C/N) parameters. The research was conducted by using biodigester in batch anaerobic operation at room temperature. Degradation process was done in 60 days. The results showed that the use of the L-AD method with TS 3.38% produced more biogas yields than using the SS-AD method. Based on the results of the research on the effect of the C/N ratio on biogas productivity using L-AD method, the optimum C/N ratio was 30. The optimum C/N ratio for biogas production from water hyacinth leaves by the SS-AD method was 32.09.

Keywords: biogas, water hyacinth, liquid anaerobic digestion (L-AD), solid-state anaerobic digestion (SS-AD), total solid (TS), carbon to nitrogen (C/N), food to microorganism (F/M)

1. Introduction

Recently, energy has become a basic need for modern society. The need of using energy was increased due to population and consumption growth and because the community used various kinds of equipment in supporting convenience in life [1]. The current global energy supply is highly dependent on fossil sources (crude oil, lignite, hard coal, natural gas). These are fossilized remains of dead plants and animals, which have been exposed to heat and pressure in the Earth's crust over hundreds of millions of years. For this reason, fossil fuels are nonrenewable resources in which reserves are being depleted much faster than the new ones being formed [2].

Indonesia as a tropical country has abundant renewable energy sources as alternative energy to replace fossil energy. One alternative energy is biogas. Biogas is the final gas product of anaerobic digestion/degradation (in an environment without oxygen) by methanogenic bacteria [1]. Biogas is very potential as the latest energy source because its methane (CH_4) content itself has a heating value of 50 MJ/kg. Methane (CH_4) has one carbon in each chain, which can produce combustion more environmentally friendly than that of the long carbon chain fuels. This matter is caused by the less amount of CO_2 produced during short carbon chain fuel combustion [3]. One of the main advantages of biogas production is the ability to transform waste material into a valuable resource, by using it as a substrate for AD [2].

Anaerobic digestion (AD) has been extensively used to convert organic waste streams from various sources, such as agricultural, industrial, and municipal solid waste, to biogas. The AD process can operate in both liquid and solid states in terms of total solid (TS) content. In general, the TS content of liquid AD (L-AD) systems ranges from 0.5 to 15%, while solid-state AD (SS-AD) systems usually operate at TS contents of higher than 15% [4].

Anaerobic digestion (AD) relies on efficient conversion of organic matter into a valuable product known as biogas, with methane (CH_4) as its main combustible constituent. The biogas can be used as energy for household cooking, lighting, heating, and other applications. The process is heavily dependent upon the mutual and syntrophic interaction of a consortium of microorganisms to break down the complex organic matter into soluble monomers such as amino acids, fatty acids, simple sugars, and glycerols. For AD process optimization, it is vital to understand these biological processes and their associated chemical reactions [5].

There are four basic stages involved in AD. These four basic stages make up the process of biogas production from various organic materials as it occurs in an anaerobic digester. These four stages are the hydrolysis, acidogenesis, acetogenesis, and methanogenesis. The AD process is characterized by the decomposition of organic matter into methane, carbon dioxide, inorganic nutrients, and compost in an anaerobic environment [6].

Many different types of anaerobic digesters are available. These vary in configuration, retention time, pre- and posttreatment requirements, and operating temperature, among other things, depending upon the principal feedstocks being treated. During AD, the breakdown of organic compounds is achieved by a combination of many types of bacteria and archaea (microbes). The biomass added to the digester is broken down into sugars, amino acids, and fatty acids (hydrolysis), fermented to produce volatile fatty acids and alcohols (acidogenesis) followed by the conversion into hydrogen, carbon dioxide, and ammonia. In addition, methanogens produce biogas from acetic acid and hydrogen [7].

The addition of biostarter can maximize biogas production. The selection of a good starter is very important to speed up the process overhaul of organic matter. Rumen fluid can be used as a good biostarter because in it there are cellulolytic and methanogenic bacteria. Cellulolytic bacteria degrade an organic material to become a substrate of methanogenic bacteria [8]. The addition of rumen fluid can also shorten the time to reach peak production of methane gas compared to substrates that are not given rumen fluid [9].

2. Biogas production from water hyacinth

Water hyacinth (*Eichhornia crassipes*) is a water plant that grows in swamps, lakes, reservoirs, and rivers and that flows calmly. The leaves of the water hyacinth are bright green, have an ovate shape, and widen with a diameter of up to 15 cm [10]. The water hyacinth (*Eichhornia crassipes*) is generally considered as a waterweed, which has become a problem that damages the environment, the system irrigation, and agriculture [11]. Water hyacinth is a type of weed that grows very fast. The growth of water hyacinth can reach 1.9% per day with a height between 0.3 and 0.5 m. Its rapid growth is felt to be very detrimental because water hyacinth plants that covered the surface of the water will cause the oxygen content to decrease [12].

Water hyacinth has attracted attention to scientists to use it as a potential biomass because its rich in nitrogen, essential nutrients, and high fermentation contents [13].

In Indonesia, most of the major lakes are also facing environmental problems such as eutrophication, sedimentation, and a decline in surface area. Indonesia has determined that 15 lakes have become a national priority to be restored and preserved [14]. Behind its beauty, Rawa Pening Lake keeps a pile of concerns. The 2667 hectare natural reservoir located in Ambarawa, Bawen, Tuntang, and Banyubiru, Semarang Regency, is currently being staked out by sedimentation, not to mention the uncontrolled growth of water hyacinth that takes up lake land. The decline in water storage capacity due to the sedimentation process results in a decrease in reservoir function and effectiveness. Rawa Pening Lake has even been included in the list of 15 critical lakes in Indonesia [15].

Rawa Pening Lake has been facing an invasion of macrophytes indicated by a massive growth of water hyacinth that covers more than 40% of the lake surface [16]. Although the water hyacinth is often used, it does not reduced. Their growth is so fast causing water hyacinth plants become into waterweeds. Water hyacinth is being utilized as a biogas raw material because it has carbohydrate and cellulose contents. Cellulose will be hydrolyzed into glucose by bacteria which will produce methane gas as biogas [10]. An image of a massive growth of water hyacinth in Rawa Pening Lake, Indonesia, is shown in **Figure 1**.

Biogas contains methane, and it is the combustion of methane which constitutes the energy component of biogas [7]. It consists mainly of methane (CH_4) and carbon dioxide (CO_2) and is formed from the anaerobic bacterial decomposition of organic compounds, i.e., without oxygen. The gases formed are the waste products of the respiration of these decomposer microorganisms, and the composition of the gases depends on the substance that is being decomposed. If the material consists of mainly carbohydrates, such as glucose and other simple sugars and high molecular compounds (polymers) such as cellulose and hemicellulose, the methane production is low. However, if the fat content is high, the methane production is likewise high [17].

Figure 1.
A massive growth of water hyacinth in Rawa Pening Lake, Central Java, Indonesia.

Biogas may be used in many different ways:

1. Combusted directly in domestic stoves for cooking or used in gas lamps for lighting

2. After minor treatment, combusted in boilers to generate heat, internal or external combustion engines to produce electricity, combined heat and power (CHP) plants to produce both heat and electricity, and tri-generation systems to provide cooling via absorption chillers in addition to heat and electricity

3. Upgraded into biomethane to be used as vehicle fuel in gas-powered vehicles; to be used in place of natural gas in industrial, commercial, and domestic uses; or to be pumped into gas grids to substitute natural gas supplied to households and businesses [7].

There were a lot of researches about biogas production that used various parameters that effected to it. These were food to microorganism (F/M) ratio, carbon to nitrogen (C/N) ratio, and total solid (TS). In the production of biogas from anaerobic digestion, the value of the food to microorganism (F/M) ratio shows the ratio between the mass of food available in the substrate and the mass of microorganisms that act as decomposers. A food to microorganism (F/M) ratio that is too small can cause microbes to be not metabolized completely, and if the value of the F/M ratio is excessive, it results in an unbalanced metabolism [18].

In addition to the organic content of the substrate, the carbon to nitrogen (C/N) ratio was stated as an important factor for the biogas process. The C/N ratio should be in the range between 10 and 30 and, as an optimal ratio, between 25 and 30 for digesters operating at full potential. When the C/N ratio is low, there is a risk of ammonia obstruction, the process of methanogenesis being more sensitive. High ratios can lead to low methane yields equivalent to the lack of nitrogen available for cell growth [19].

According to Brown and Li (2013) in the production of biogas from biomass raw materials, lignocellulose is appropriate to be produced from using the SS-AD method because lignocellulosic biomass has a total solid concentration of >15% and has low moisture content. According to Malik (2006) water hyacinth contains 95% water and consists of networks that are hollow, and this is the reason why L-AD method is well applied to water hyacinth because of its TS content which is relatively low [20].

Some researches about biogas production of water hyacinth have been done by students of the Environmental Engineering Diponegoro University. The researches were about biogas production from water hyacinth using liquid anaerobic digestion (L-AD) and solid-state anaerobic digestion (SS-AD). The part that was used from water hyacinth was the leaves.

2.1 Measurement methods

2.1.1 Preliminary methods

Preliminary methods were conducted before doing the main researches to know about the contents of each component that will be used. Various parameters will be used in biogas researches.

2.1.1.1 Total solid

According to the American Public Health Association (APHA) standard method, the formula for total solid content can be seen in Eq. (1):

$$\left[\frac{W3 - W1}{(W2 - W1)}\right] \times 100\% \tag{1}$$

Description: $W1$, cup weight; $W2$, cup weight and sample weight; $W3$, cup weight and sample weight after being ovened.

2.1.1.2 Measuring C-organic content using the Walkley and Black method

The procedure carried out in this test was taken from several references, namely, Black (1965); Graham (1948); Page et al. (1982); Rayment et al. (1992) in Sulaeman et al. (2005) "Technical Guidelines for Soil, Plant, Water, and Chemical Chemical Analysis of Soil Research Institute Indonesian Ministry of Agriculture." With the following method of work, 0.500 g soil of size <0.5 mm was weighed and put in a 100 ml volumetric flask. 5 ml of $K_2Cr_2O_7$ 1 N was added and then mixed. 7.5 ml of concentrated H_2SO_4 was added, mixed, and let to sit for 30 min. Diluted with ion free water, the mixture was allowed to cool and squeeze. In the next day, absorbance of the clear solution was measured with a spectrophotometer at a wavelength of 561 nm. As a comparison standard, 0 and 250 ppm were made, by piping 0 and 5 ml of the 5000 ppm standard solution into a 100 ml volumetric flask with the same treatment as the working sample [21].

2.1.1.3 Measuring N-total levels using a spectrophotometer

In this test, the procedure was taken from several references, namely, Black, (1965); Page et al. (1982); Burt (2004); and Lisle et al. (1990) in Sulaeman et al. (2005) "Technical Guidelines for Soil, Plant Chemical Analysis, Water and Fertilizers, Indonesian Ministry of Agriculture Soil Research Institute." This test was divided into two stages: the destruction stage and the measurement stage [21].

2.1.2 Biogas measurement

Samples that had been researched in preliminary methods could be inserted into the reactor and mixed with other components that are related; then the reactor must be sealed in order to obtain anaerobic digestion. During the treatment process, the volume of biogas production was observed in an interval of 2 days throughout 60 days.

To find out the amount of biogas, place the reversed cylinder glass in the container that is filled with water (reversed cylinder glass must be filled with water). Place the plastic tube into the reversed cylinder glass. Record the initial volume from it. Open the clip that clipped the plastic tube (the clip's function was to avoid the oxygen entered into the digester). The biogas will go out through the plastic tube and will make the water volume to decrease. Record the final volume. Lastly, record the biogas volume by counting the water level difference. The digester is shown in **Figure 2**.

2.2 Liquid anaerobic digestion (L-AD) method

2.2.1 The effect of total solid (TS)

TS content of liquid anaerobic digestion (L-AD) systems ranges from 0.5 to 15% [4]. The research about "Biogas Production from Water Hyacinth (*Eichhornia crassipes*): The Effect of F/M Ratio" [22] was conducted to know about the effect of F/M ratio to biogas production from water hyacinth leaves using the liquid anaerobic digestion (L-AD) method. In biogas production anaerobically, the value of F/M shows a comparison between the amount of substrate that is contained in waste (medium)

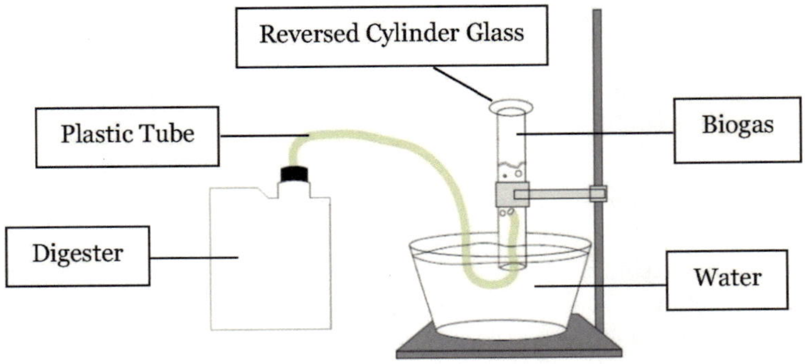

Figure 2.
Schematic diagram of series laboratory batch assessment of L-AD and SS-AD [23].

F/M ratio	Initial total solid (%)	Cow rumen fluid volume (ml)	Final total solid (%)
39.76	13.52	50	10.82
20.03	13.52	100	9.06
13.32	13.52	150	7.73
10.01	13.52	200	6.76

Table 1.
Initial and final total solid of water hyacinth leaves.

and the amount of microorganism used [18]. The variation of F/M ratio depends on the existence of rumen volume variation and total solid from each materials.

The main substrate used in the research [22] was water hyacinth leaves as much as 200 g. The initial total solid of water hyacinth leaves that has been calculated using (Eq. (1)) was 13.52. When it is combined with a different cow rumen fluid volume in each reactor, the total solid of water hyacinth leaves will be changed. To find out the F/M ratio, a comparison of the total solid of water hyacinth leaves with cow rumen fluid was multiplied by the weight and volume of each ingredient. The data is shown in **Table 1**.

After the research had been done, results show that the biogas production with F/M ratio of 10.01 and TS of 6.76% produced more biogas in the amount of 127.071 ml/g TS. **Figure 3** shows the graphic of cumulative biogas yield.

A research has also been conducted [20] using water hyacinth leaves as much as 200 g as the main substrate combined with water and rumen fluid. The combination is shown in **Table 2**.

In this study to get a low total solid content, rumen was added to the first variable, and water and rumen were added to the second variable. This is consistent with the research conducted by Astuti (2013) which states that the stuffing material must contain about 6–9% dry matter. This situation can be achieved by dilution [20]. From the graphic below, the final result of biogas production with a TS variable of 6.76% was 177.33 ml/g TS and for a TS variable of 3.38% was 369 ml/g TS. The graphic of cumulative biogas yield/TS is shown in **Figure 4**.

2.2.2 The effect of C/N ratio

In addition to total solid and F/M ratio, biogas production is also affected by carbon to nitrogen (C/N) ratio. Various C/N ratio researches have been done [23] by adding organic compound that contained high nitrogen such as urea.

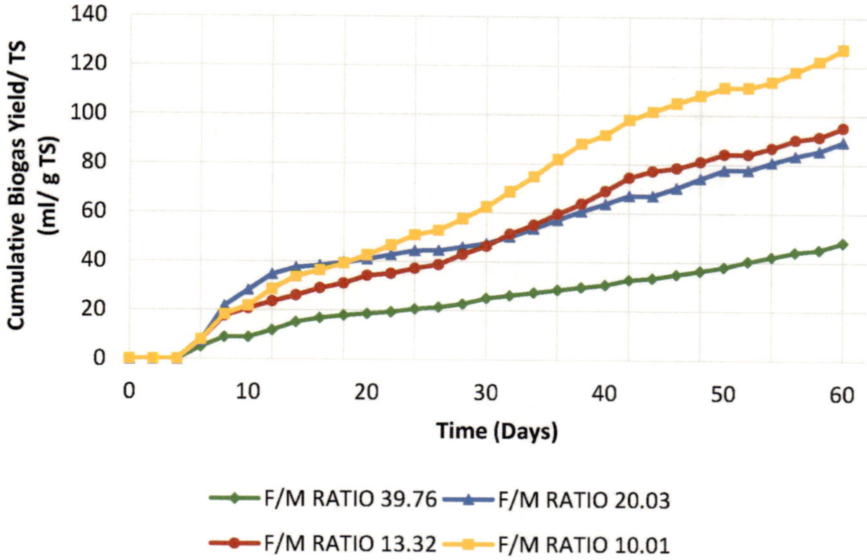

Figure 3.
Cumulative biogas yield per g TS based on F/M ratio.

Initial total solid (%)	Cow rumen fluid volume (ml)	Water volume (ml)	Final total solid (%)
13.52	150	—	6.76
13.52	150	300	3.38

Table 2.
Initial and final total solid of water hyacinth leaves.

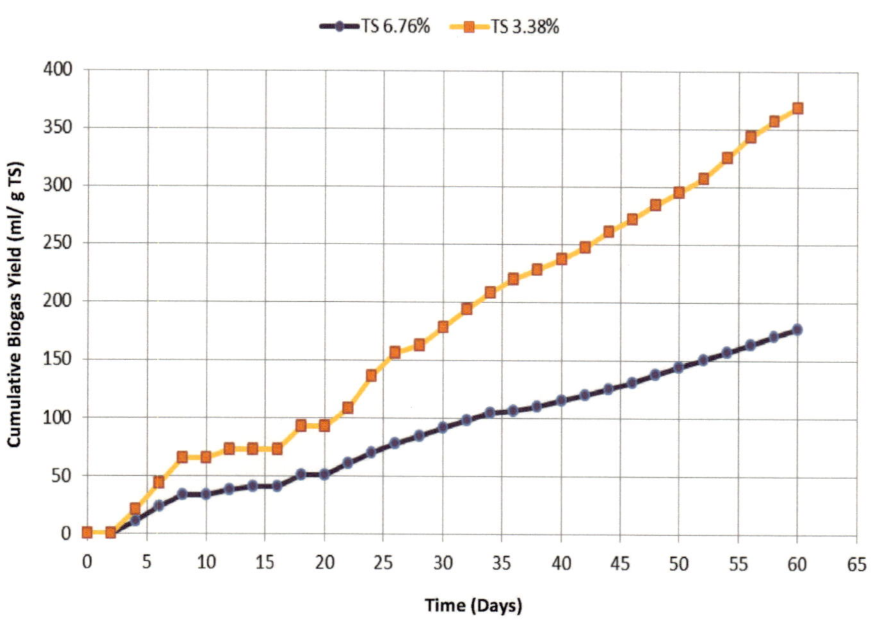

Figure 4.
Cumulative biogas yield per gram TS.

In the variation of N elements, C/N ratios of 20, 25, 30, and 35 were produced. The material components of the variables are shown in **Table 3**.

The equation for C/N ratio had a real influence on the production of biogas with water hyacinth leaves as a raw material. A variation of C/N ratio of 30 gave the best rate of biogas production among other C/N ratio variables, with biogas yield generated at 191,423 ml/g TS [23]. The result of biogas cumulative yield with C/N ratio can be seen in **Figure 5**.

2.3 Solid-state anaerobic digestion (SS-AD)

2.3.1 The effect of total solid (TS)

Solid-state anaerobic digestion (SS-AD) systems usually operate at TS contents of higher than 15% [4]. A research has been conducted [20] about the effect of

Variable	Cow rumen fluid volume (ml)	Water hyacinth leaves (g)	Urea (g)/C/N ratio			
			20	25	30	35
1	200	200	4,9			
2	200	200	4,9			
3	200	200		3,4		
4	200	200		3,4		
5	200	200			3,0	
6	200	200			3,0	
7	200	200				2,5
8	200	200				2,5

Table 3.
Research material needs.

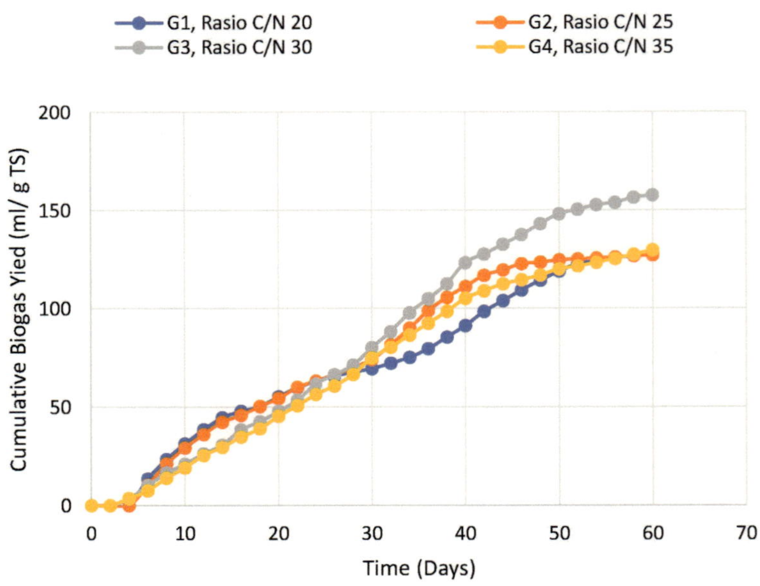

Figure 5.
Cumulative biogas yield per g TS based on C/N ratio.

solid-state anaerobic digestion (SS-AD) on biogas production using water hyacinth leaves. In this study to make the total solid content increased, drying method was used. The water hyacinth leaves from Rawa Pening Lake have an initial total solid content of 13.52%. The first variable was dried for 2 days, and the second variable was dried for 1 day. After that, the water hyacinth leaves that have been dried in the sun were examined for their total solid content using Eq. (1). For the first variable, the total solid content was 48.26% and for the second variable was 36.36%. After that the water hyacinth leaves that have been dried in the sun were added to the cow rumen in a ratio of 1:1. In studies using the SS-AD method, no additional water was given [20]. The combination of the variables is shown in **Table 4**.

The variable with TS of 24.34% produced biogas with a total of 34.79 ml/g TS, and the variable with TS of 17.67% obtained 52.98 m/g TS. The result is shown in **Figure 6**.

Further research had been conducted [24] to know about the optimization of total solid (TS) and carbon to nitrogen (C/N) ratio of biogas production from water hyacinth leaves by adding microbial consortium as much as 3%, 6%, and 9%. Meanwhile the total solid contents from the research were 15%, 27.5%, and 40%. And the C/N ratios were 20, 35, and 50. To get the optimum conditions, calculation had been done by the central composite design method with the following variables in **Table 5**.

Variations of variable values in each reactor were obtained using Statistica software as shown in **Table 6**.

Initial TS (%)	Final TS after being dried in the sun (%)	Cow rumen fluid (ml)	Final TS (%)
13.52	48.26	150	24.13
13.52	36.36	150	17.67

Table 4.
Initial and final total solid of water hyacinth leaves.

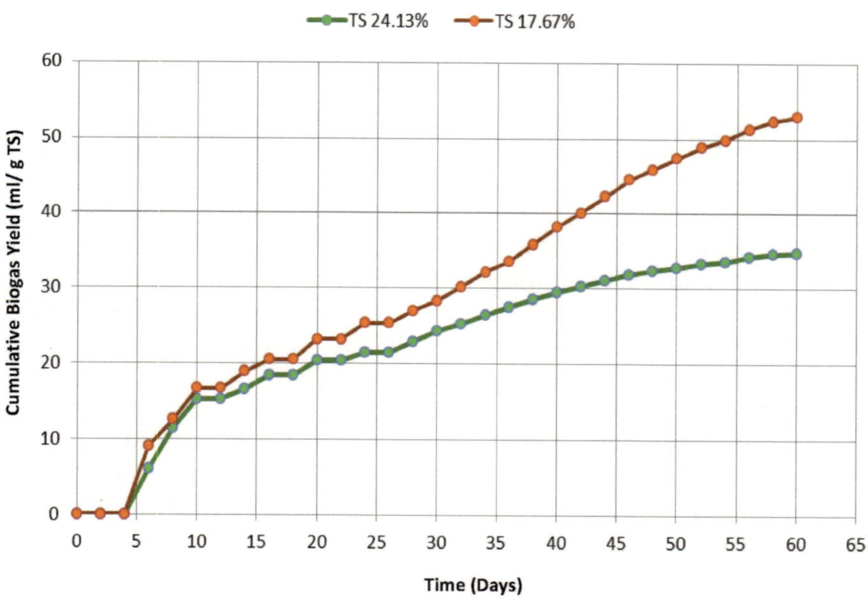

Figure 6.
Cumulative biogas yield per gram TS.

Parameter	−1.682	−1	0	+1	+1.682
Microbial consortium (%)	0.7085	3	6	9	11.2915
C/N ratio	8.54249	20	35	50	61.45751
Total solid (%)	5.45207	15	27.5	40	49.54793

Table 5.
Variable values in the central composite design.

Reactors	Total solid	C/N ratio	Microbial consortium
1	15	20	3
2	15	20	9
3	15	50	3
4	15	50	9
5	40	20	3
6	40	20	9
7	40	50	3
8	40	50	9
9	27.5	35	6
10	5.45207	35	6
11	49.54793	35	6
12	27.5	8.54249	6
13	27.5	61.45751	6
14	27.5	35	0.70850
15	27.5	35	11.29150
16	27.5	35	6
17	27.5		

Table 6.
Variable values in experiments using central composite design.

In this SS-AD method, variations in the total solid concentration used were 15%, 27.5%, and 40%. The total solid for each reactor was adjusted to the total solid of the water hyacinth. Water and rumen were added to regulate the total solid in each of the reactors. **Figures 7–9** show the graphs of biogas results produced at certain reactors which were compared between reactors with the same C/N and microbial consortium ratio values against different TS values [24].

The graph in **Figure 7** shows the production of biogas produced from Reactor 1 and Reactor 5 where the reactors had the same concentration variations for the same C/N ratio and microbial consortium variables, with the lowest value of each variation of 20 for the C/N ratio and 3% for the concentration of the microbial consortium. The difference was in the total solid concentration (**Table 6**). Based on the graph in **Figure 7**, the total cumulative biogas production for Reactor 1 was 27.367 ml/g TS while for Reactor 5 was 5.1 ml/g TS. Reactor 1 with a lower TS which was 15% produced biogas with a greater total than that of the Reactor 5 with TS of 40% [24].

Reactors 4 and 8 had varying concentrations for the same C/N and microbial consortium variables (**Table 6**), namely, a C/N ratio of 50 and a microbial consortium concentration of 9%. Both of these variation values are the highest values

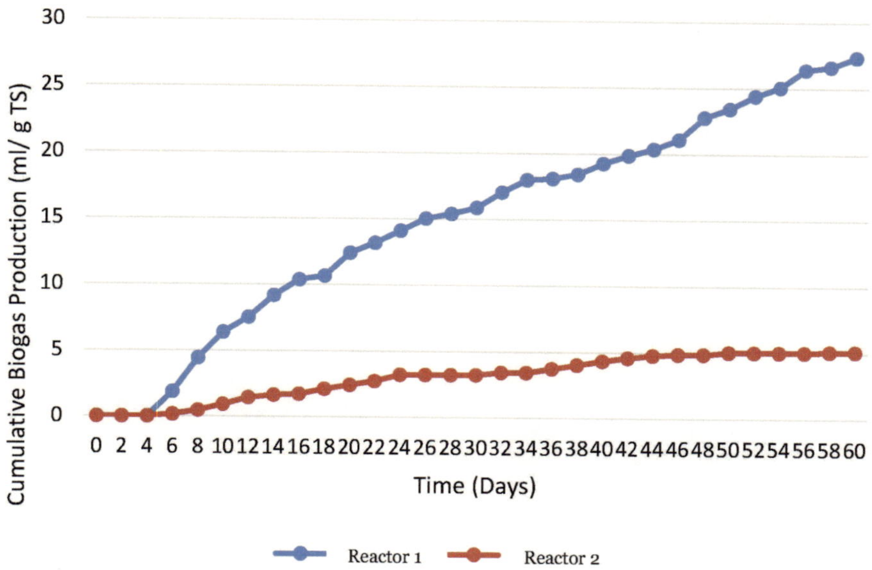

Figure 7.
Effect of TS on biogas production (Reactors 1 and 5).

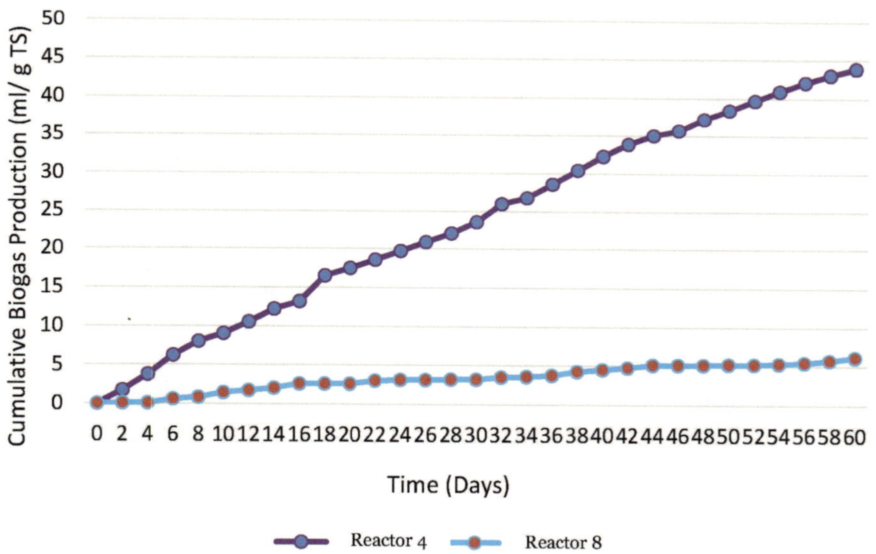

Figure 8.
Effect of TS on biogas production (Reactors 4 and 8).

among the range of values for these variables. The TS concentrations in Reactors 4 and 8 were 15% (lowest value) and 40% (highest value). Reactor 4 with a lower TS value of 15% produces more biogas production than Reactor 8 with a higher TS value (40%). **Figure 8** shows that the total biogas production for Reactor 4 was 43.87 ml/g TS and for Reactor 8 was 6.15 ml/g TS [24].

The biogas production graph in **Figure 9** came from a reactor with a C/N ratio of 35 and a microbial consortium concentration of 6% (**Table 6**). This value was the middle value of the variation of concentration for each of these variables. Biogas production varies in each of the reactors. It can be seen in **Figure 9** that the total biogas production for Reactors 9, 10, and 11 was 22.65 ml/g TS, 87.85 ml/g TS, and 10.09 ml/g TS. Reactor 10 with TS concentration of 5.45%, C/N ratio of 35, and

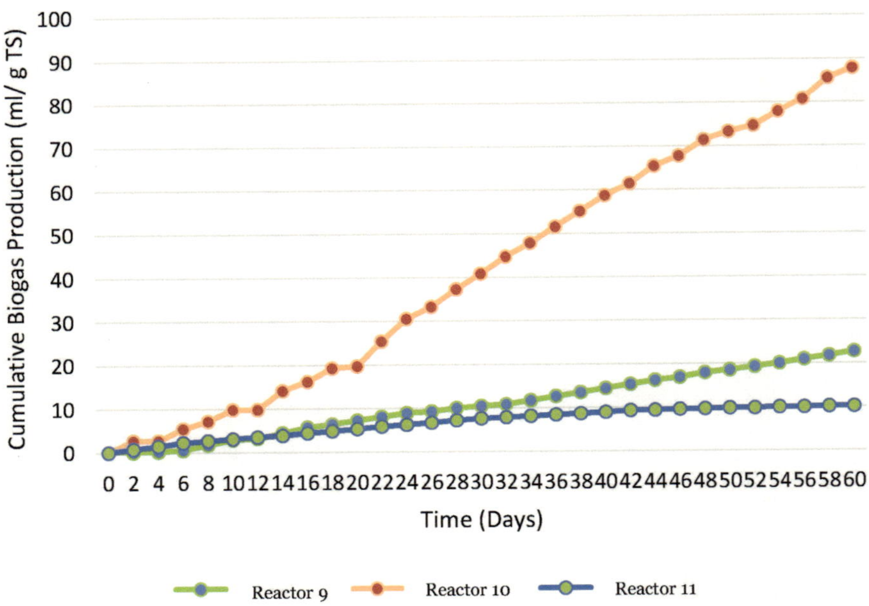

Figure 9.
Effect of TS on biogas production (Reactors 9, 10, and 11).

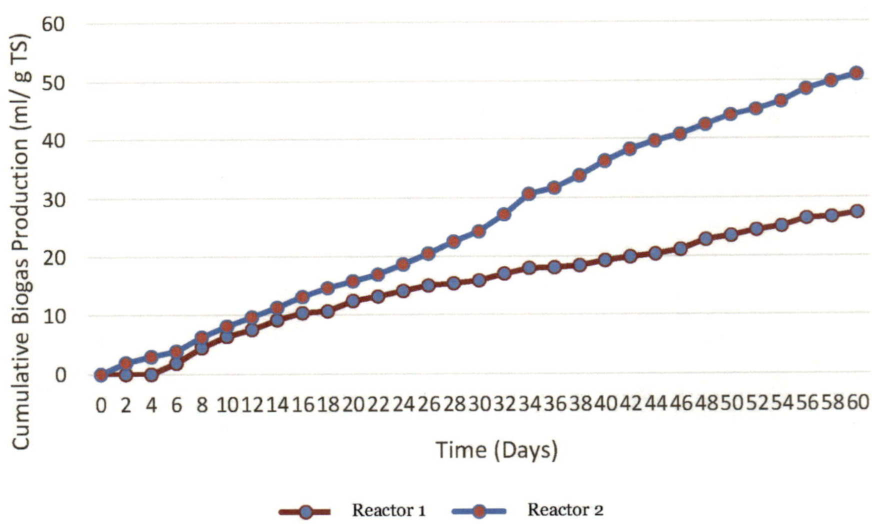

Figure 10.
Effect of C/N ratio on biogas production (Reactors 1 and 3).

microbial consortium concentration of 6% produces the largest biogas production when among Reactors 10 and 11 [24].

2.3.2 The effect of C/N ratio

The variations in the C/N ratio used in this study were 20, 35, and 50. First the C/N ratios of the water hyacinth leaves were tested. To obtain variations in the concentration of the C/N ratio as determined, urea was used to adjust the N value of the water hyacinth leaves [24].

Different C/N ratios were tested with the same total solid and microbial consortium concentration in Reactor 1 and Reactor 3 (**Table 6**). Reactor 1 with a C/N ratio of 20, total solid of 15%, and microbial consortium concentration of 3% produced a total biogas of 27.37 ml/g TS. Reactor 3 with a C/N ratio of 50, total solid of 15%, and microbial consortium concentration of 3% produced biogas with a total of 51 ml/g TS. Reactor 3 with a higher C/N ratio of 50 produced more biogas volume than the Reactor 1 with a C/N ratio of 20. The graph is shown in **Figure 10** [24].

The graph in **Figure 11** was a biogas graph produced from Reactors 6 and 8. The concentrations of the total solid and microbial consortium variables in Reactor 6 were the same as those in Reactor 8 which were 40% and 9%, respectively (**Table 6**). The C/N ratio of Reactor 6 was 20, while Reactor 8 is 50. For the total biogas production produced, based on the graph in **Figure 11**, it can be seen that Reactor 6 has a higher biogas than that of the Reactor 8 which was 13.14 ml/g TS for Reactor 6 and 6.15 ml/g TS for Reactor 8. Thus, reactors with lower C/N ratios produce higher biogas under conditions of total solid concentration of 40% and microbial consortium of 9% [24].

The graph in **Figure 12** was taken from the calculation of biogas production produced in Reactors 9, 12, and 13. The reactors have the same total solid concentration and

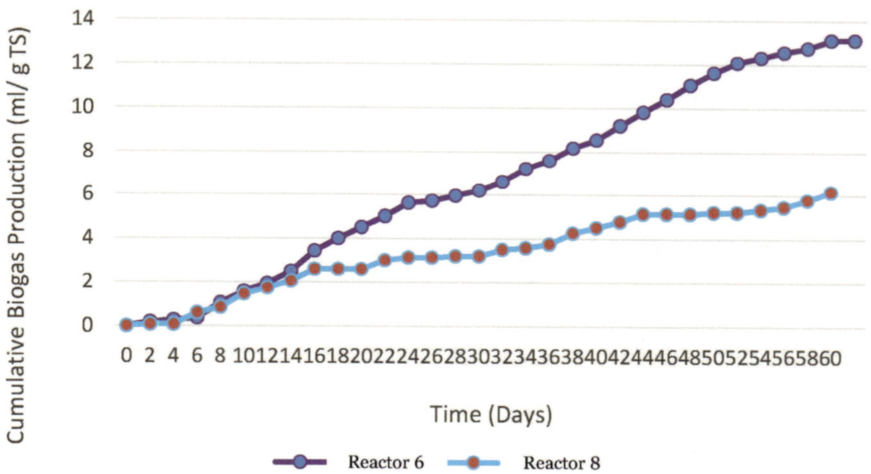

Figure 11.
Effect of C/N ratio on biogas production (Reactors 6 and 8).

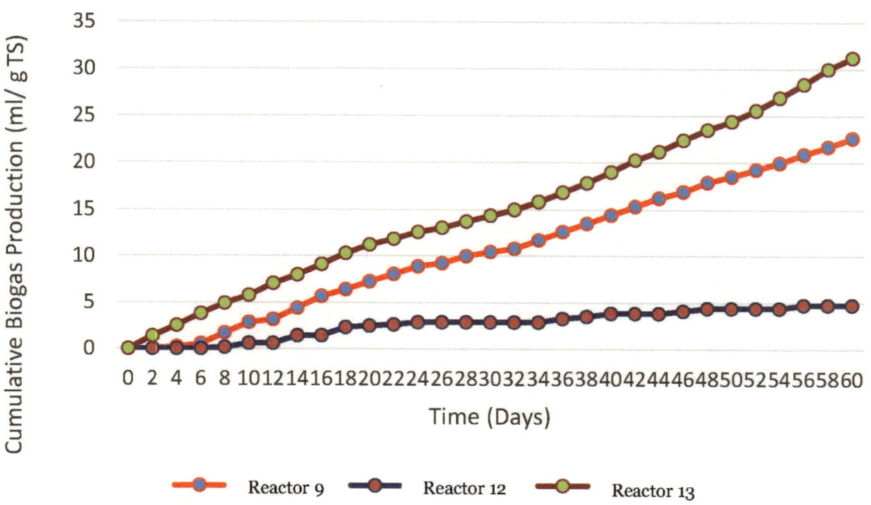

Figure 12.
Effect of C/N ratio on biogas production (Reactors 9, 12, and 13).

microbial consortium (**Table 6**) of 27.5% and 6%, respectively, with a C/N ratio different from Reactor 9 with a C/N ratio of 35, Reactor 12 with a C/N ratio of 8.54, and Reactor 13 with a C/N ratio of 61.45. The total biogas production in Reactor 9 was 22.65 ml/g TS, whereas in Reactor 12, the total biogas production was 4.76 ml/g TS. For Reactor 13, the total biogas production was 31.24 ml/g TS. The volume of biogas production in Reactor 13 was greater than the volume of biogas production in Reactors 9 and 12 [24].

3. Conclusions

Water hyacinth was considered as waterweed, which has become a problem that damaged the environment, the system irrigation, and agriculture. Water hyacinth leaves that contained cellulose, nitrogen, essential nutrients, and high fermentation contents can be used for biogas production. The use of the L-AD method with TS 3.38% produced the most biogas yields than using the SS-AD method with TS 24.13 and TS 17.67 or the L-AD method with TS 6.76%, with the amount of biogas yield for TS 3.38% was 369 ml/g TS.

Based on the results of research on the effect of the C/N ratio on biogas productivity using L-AD method, the optimum C/N ratio was found in the C/N ratio 30 with the resulting biogas yield of 157.544 ml/g TS. The optimum C/N ratio for biogas production from water hyacinth leaves using the solid-state anaerobic digestion method was 32.09.

Acknowledgements

We would like to say thank you to the Diponegoro University for the funding of this research under the Research Professorship Program (RPP) (2017).

Notes/thanks/other declarations

Praise and gratitude for the presence of Allah SWT who has bestowed His mercy and grace so that the author can complete the part of the book with the title Biogas Production from Water Hyacinth chapter. In completing this chapter, we would like to thank the Environmental Engineering Diponegoro University and the water hyacinth biogas research team for their guidance and support. We hope the research that has been done can bring benefits to all people.

Appendices and nomenclature

AD	anaerobic digestion
	biogas
C/N	carbon to nitrogen
	cumulative biogas yield
F/M	food to microorganism
	Rawa Pening Lake
L-AD	liquid anaerobic digestion
	methane
SS-AD	solid-state anaerobic digestion
TS	total solid
	water hyacinth

Author details

Winardi Dwi Nugraha, Syafrudin and Lathifah Laksmi Pradita*
Diponegoro University, Semarang, Indonesia

*Address all correspondence to: lathifahlp@gmail.com

References

[1] Saputri YF, Yuwono T, Mahmudsyah S. Pemanfaatan Kotoran Sapi untuk Bahan Bakar PLT Biogas 80 KW di Desa Babadan Kecamatan Ngajum Malang. Jurnal Teknik Pomits [Internet]. 2014;**1**(1):1-6. Available from: http://digilib.its.ac.id/public/ITS-paper-33637-2212105070-Paper.pdf

[2] Al Seadi T, Rutz D, Prassl H, Kottner M, Finsterwalder T, Volk S, et al. Biogas Handbook [Internet]. Esbjerg: University of Southern Denmark Esbjerg; 2008. p. 10. Available from: https://www.;emvigbiogas.com/BiogasHandbook.pdf

[3] Suyitno, Sujono A, Dharmanto. Teknologi Biogas: Pembuatan, Operasional dan Pemanfaatan. Yogyakarta: Graha Ilmu; 2010. p. 2

[4] Zhu J, Zheng Y, Xu F, Li Y. Solid-state anaerobic co-digestion of hay and soybean processing waste for biogas production. Bioresource Technology [Internet]. 2014;**154**:240-247. Available from: www.elsevier.com/locate/biortech

[5] Anukam A, Mohammadi A, Naqvi M, Granstrom K. A review of the chemistry of anaerobic digestion: Methods of accelerating and optimizing process efficiency. Processes [Internet]. 2019;**7**(504). DOI: 10.3390/pr7080504. Available from: www.mdpi.com/journal/processes

[6] Sawyerr N, Trois C, Workneh T, Okudoh V. An overview of biogas production: Fundamentals, applications and future research. International Journal of Energy Economics and Policy [Internet]. 2019;**9**(2):105-116. DOI: 10.32479/ijeep.7375. Available from: https://www.econjournals.com/index.php/ijeep/article/view/7375

[7] Jain S. Global Potential of Biogas [Internet]. London: World Biogas Association; 2019. p. 56. Available from: https://www.worldbiogasassociation.org/wp-content/uploads/2019/09/WBA-globalreport-56ppa4_digital-Sept-2019.pdf

[8] Ningsih SS, Ahda Y, Handayani D. Pengaruh Penambahan beberapa Cairan Rumen Terhadap Produksi Biogas dari Kotoran Sapi. Biospecies [Internet]. 2014;**7**(2):34-42. Available from: https://online-journal.unja.ac.id/biospecies/article/view/1734

[9] Susilowati E. Uji Potensi Pemanfaatan Cairan Rumen Sapi untuk Meningkatkan Kecepatan Produksi Biogas dan Konsentrasi Gas Metan dalam Biogas [thesis]. Yogyakarta: Gadjah Mada University; 2009

[10] Darnengsih, Nurjannah, Ifa L. Pengaruh Perbandingan Bahan Baku terhadap Konsentrasi Biogas dari Eceng Gondok dengan Menggunakan Starter Kotoran Sapi. Journal of Chemical Process Engineering [Internet]. 2016;**01**(01). Available from: https://www.academia.edu/37439576/PENGARUH_PERBANDINGAN_BAHAN_BAKU_TERHADAP_KONSENTRASI_BIOGAS_DARI_ECENG_GONDOK_DENGAN_MENGGUNAKAN_STARTER_KOTORAN_SAPI

[11] Chuang YS, Lay CH, Sen B, Chen CC, Gopalakrishnan K, Wu JH, et al. Biohydrogen and biomethane from water Hyacinth (*Eichhornia crassipes*) fermentation: Effects of substrate concentration and incubation temperature. International Journal of Hydrogen Energy [Internet]. 2011;**36**:14195-14203. DOI: 10.1016/j.ijhydene.2011.04.188

[12] Yonathan A, Prasetya AR, Pramudono B. Produksi Biogas dari Eceng Gondok (Eicchornia Crassipes): Kajian Konsistensi dan pH terhadap Biogas Dihasilkan. Jurnal teknologi Kimia

dan Industri [Internet]. 2013;**2**(2):211-215. Available from: http://ejournal-s1.undip.ac.id/index.php/jtki

[13] Patil JH, AntonyRaj M, Shankar BB, Shetty MK, Kumar BPP. Anaerobic co-digestion of water hyacinth and sheep waste. Energy Procedia. 2014;**52**:572-578. DOI: 10.1016/j.egypro.2014.07.112

[14] State Minister for the Environment. Profil 15 Danau Prioritas Indonesia. 2011. pp. 148

[15] Sedimentasi Makin Parah, Rawa Pening Terancam Punah [Internet]. 2019. Available from: https://www.jawapos.com/jpg-today/20/02/2019/sedimentasi-makin-parah-rawa-pening-terancam-punah/ [Accessed: 10 January 2020]

[16] Suprobowati TS, Hadisusanto S, Gell P, Zawadski A. The diatom stratigraphy of Rawapening Lake, implying eutrophication history. American Journal of Environmental Sciences [Internet]. 2012;**8**(3):334-344. Available from: https://www.researchgate.net/publication/276280342_The_Diatom_Stratigraphy_of_Rawapening_Lake_Implying_Eutrophication_History

[17] Jorgensen PJ. Biogas—Green Energy. Aarhus: Faculty of Agricultural Sciences, Aarhus University; 2009

[18] Hadiyarto A, Budiyono B, Johari S, Hutama I, Hasyim W. The effect of F/M ratio to the anaerobic decomposition of biogas production from fish offal waste. Waste Technology. 2015;**3**(2):58-62. DOI: 10.12777/wastech.3.2.58-61

[19] Teghammar A. Biogas production from lignocelluloses: Evaluation [thesis]. Göteborg: Chalmers University of Technology; 2013

[20] Natalyn F. Studi Pengaruh Metode L-AD (Liquid anaerobic digestion) dan SS-AD (solid state anaerobic digestion) Terhadap Produksi Biogas Daun Eceng Gondok (Eichhornia crassipes) [thesis]. Semarang: Diponegoro University; 2017

[21] Febrina M. Optimasi Produksi Biogas Metode Liquid anaerobic digestion (L-AD) dari Limbah Eceng Gondok Dengan RSM (response surface methodology) [thesis]. Semarang: Diponegoro University; 2019

[22] Nugraha WD, Syafrudin PLL, Matin HHA, Budiyono. Biogas production from water hyacinth (*Eichhornia crassipes*): The effect of F/M ratio. In: Proceeding of the International Conference on Future Environment and Energy (ICFEE '18); 10-12 January 2018; Thailand. Hong Kong: ICFEE; 2018. pp. 129-134. DOI: 10.1088/1755-1315/150/1/012019

[23] Munkar G. Pengaruh C/N ratio pada Produksi Biogas dari Daun Eceng Gondok dengan Metode liquid anaerobic digestion (L-AD) [thesis]. Semarang: Diponegoro University; 2017

[24] Senduk AT. Kajian Produksi Biogas Metode solid-state anaerobic digestion (SS-AD) dari Daun Eceng Gondok Dengan response surface methodology (RSM) [thesis]. Semarang: Diponegoro University; 2018

Chapter 4

Solid-State Anaerobic Microbial Ensilage Pretreatment

Xu Yang

Abstract

Pretreatment technology has become the main bottleneck restricting the development of biogas. This chapter provides an overview of recent studies on solid-state microbial ensilage pretreatment for the production of biogas with wastes. The goal of microbial ensilage pretreatment is to maximize the production of lactic acid, thereby reducing the pH value and establishing an environment that is not suitable for the growth of harmful organisms. The use of various additives, especially lactic acid bacteria, is the main factor to ensure the success of anaerobic pretreatment. Sensory evaluation is carried out by observing the smell, structure, and color of silage to judge the quality of silage. The pH values, ammonia nitrogen, and organic acids (lactic-, acetic-, propionic-, and butyric acid) are used as reference values to determine the fermentation quality of silage. An overall comparison of the effectiveness of microbial ensilage with aerobic microbial pretreatment for biogas production is also discussed. Finally, the research on solid-state anaerobic microbial silage pretreatment in biogas conversion is summarized. The combined anaerobic digestion method with different pretreated materials will be the future development direction due to its advantages.

Keywords: biogas, biomass, ensiling, pretreatment, wet storage

1. Introduction

At present, a large number of organic wastes (such as straw, animal manure, excess sludge, and other wastes) are produced in industry, agriculture, and aquaculture every year. How to deal with organic wastes sustainably has become a global challenge. Anaerobic digestion (AD) technology for the stable utilization of organic wastes (mineralizing volatile solids and reducing pathogens) has been applied all over the world. AD is a biological process in which organic matter is decomposed by an assortment of microbes under oxygen-free conditions and produces biogas (about 50–60% CH_4 and 25–30% CO_2) [1]. Physical approaches, mainly compression and liquefaction, have been commercially applied to upgrade biogas to bio-compressed natural gas (CNG) and liquefied biogas (LBG) [2]. Meanwhile, the nutrient-rich biogas slurry is commonly used as an organic fertilizer [3]. At present, AD with various wastes to produce biogas is the main method to solve the problems of energy shortage and environmental pollution [4]. During the development of AD engineering, the difficulty faced is how to improve the conversion efficiency of waste. Various wastes have unstable AD and low conversion efficiency due to structural composition and nutritional imbalance. The agricultural waste is closely bound together by the covalent bonds among its main components cellulose, hemicellulose, and lignin, which greatly hinders the degradation of carbohydrates [5]. Especially for corn stover, it found that the index changed after the straw was dried, which was very unfavorable for biogas fermentation and other bioenergy conversion methods. For example, the lignin content is almost doubled, which seriously hinders the degradation of cellulose and hemicellulose in the process of biochemical transformation. AD will lead to problems such as slow start of fermentation, long fermentation time, and low gas production rate. Therefore, pretreatment becomes a necessary step for AD of lignocellulose. In the field of biomass transformation, the digestibility of cellulose is affected by many factors, such as hemicellulose content, lignin characteristics (content and distribution), matrix-specific surface area and porosity, cellulose crystallinity, and cell wall thickness [6]. The purpose of pretreatment is to remove or destroy the complex structure between cellulose-hemicellulose-lignin, improve the effective contact between cellulase and biomass, and then increase the rate of enzymatic hydrolysis [7, 8]. So far, scientific researchers in various countries have developed many promising pretreatment technologies, but most of them are chemical methods that require the addition of chemical reagents, such as acid, alkali, ammonia, organic solvents, or ionic liquids, and require certain high-temperature conditions and corresponding special reaction equipment [9–12]. The effect of different pretreatment methods is different, resulting in a great difference in the final gas production situation.

Biological pretreatment is to destroy the cell wall structure of biomass by the metabolic activity of microorganisms, which has the characteristics of mild use conditions, low cost, and has great potential in the field of biomass pretreatment [13]. Compared with chemical methods, biological methods do not need to consume a large amount of energy and recover chemical reagents, nor produce harmful inhibitors in the reaction system [14]. In parallel, microbes have evolved mechanisms, including cellulose-degrading enzymes, to degrade plant cell walls to access the plants' nutritious sugars. Fungi (aerobic pretreatment) use degradative enzymes called cellulases, whereas in bacteria (anaerobic pretreatment), multiple enzymes self-assemble into a complex called the cellulosome [15]. Both biological pretreatment methods have their advantages and disadvantages. Therefore, the main purpose of the chapter was to provide an overview of recent studies on solid-state microbial pretreatment for the production of biogas with wastes, focusing on the steps involved in the anaerobic ensilage formation with microbial ensilage pretreatment, additives, and quality evaluation of the silage.

2. Aerobic microbial pretreatment

From both economic and environmental perspectives, fungal pretreatment with lignin-degrading microorganisms, preliminary white-rot fungi, has received renewed interest as an alternative to thermal/chemical pretreatment for biogas production [16]. The degradation of lignin by white-rot fungus is a process of biological oxidation. Under suitable conditions, the mycelium of white-rot fungus first secretes super fiber oxidase to dissolve the wax on the plant surface, and then the mycelium enters the plant interior and secretes the enzyme system to degrade lignin to complete the degradation of lignin. The results show that white-rot fungus can not only degrade lignin but also protect cellulose from damage, so as to improve the quality of lignocellulose and make it easier to be degraded by anaerobic bacteria. Ghosh and Bhattacharyya used *Phanerochaete chrysosporium* and *Polyporus ostreiformis* to treat rice straw, by which biogas and methane production was increased by about 34.73 and 46.19% in treated straw, respectively [17]. Taniguchi et al.

compared the gas production effect of four kinds of white-rot fungi (*Trametes versicolor*, *Phanerochaete chrysosporiu*, *Ceriporiopsis subvermispora*, and *Pleurotus ostreatus*) on the pretreatment of rice straw. It was found that *P. ostreatus* selectively degraded the lignin fraction of rice straw rather than the holocellulose component [18]. Zhong et al. analyzed the pretreatment effect of *Pleurotus florida* on anaerobic fermentation of corn straw to produce biogas. AD experiments showed that the biogas productivity was increased by all the pretreatments and that the biogas production after NaOH pretreatment would be 20.07 and 16.58% higher than the raw corn straw and biologically pretreated ones, respectively [19]. Mackulak et al. used *Auricularia auricula-judae* to pretreat sweet chestnut leaves and hay at 37°C for 4–5 weeks, which had a 15% increase in biogas production compared with the untreated samples [20].

The advantages of this technology over thermochemical pretreatments include simple techniques, low energy requirements, no or reduced output of waste streams, reduced downstream processing costs, and no or reduced inhibitors to biogas fermentation. Despite the advantages, substantial holocellulose loss and long pretreatment time are the major issues associated with fungal pretreatment. The growth cycle of microorganism is long, such as white-rot fungus, which usually takes 7–15 days generally. White-rot fungi are sensitive to temperature, which grow faster at 26–32°C, and their growth will be inhibited if the temperature is too high or too low. Meanwhile, compared with chemical and physical methods, the biodegradation efficiency of lignocellulose by biological method is not high, while the pH value and material composition will also affect the growth of microorganisms. For example, white-rot fungus grows better under the condition of partial acid, and its growth and enzyme activity will be hindered with the increase in pH value. In addition, there are a few kinds of microorganisms that could degrade lignin, while the low enzyme activity is also an important factor limiting its application.

3. Anaerobic microbial pretreatment

Microbial ensilage under anaerobic conditions can be a good way to avoid the above problems of aerobic pretreatment, which become a reliable method for long-term storage of lignocellulosic biomass (LCB) [21]. Its basic principle is to use the anaerobic fermentation of lactic acid bacteria (LAB) in closed conditions to convert the soluble carbohydrates to organic acids, which inhibit the growth of detrimental microorganisms by a strong drop in pH to values between 3 and 4 [22]. After a year of microbial ensilage, the dry matter (DM) loss of raw materials is as low as 1–5%, while the digestibility is higher than that of dry storage [23], mainly because the degradation of non-structural carbohydrates reduces the biological resistance of LCB [24]. Microbial ensilage can solve the problems of straw collection, preservation, and pretreatment in large scale. Compared with the chemical methods, the technology does not need to consume a lot of energy and recover chemical reagents, nor produce harmful inhibitors in the reaction system. To date, the silage remains one of the main methods for the efficient utilization of LCB.

3.1 The steps involved in the anaerobic ensilage formation

High-quality silage begins with harvesting at an appropriate stage of maturity to maximize the nutrient production. Next, the following management methods are essential for the successful fermentation and preservation of product (**Figure 1**).

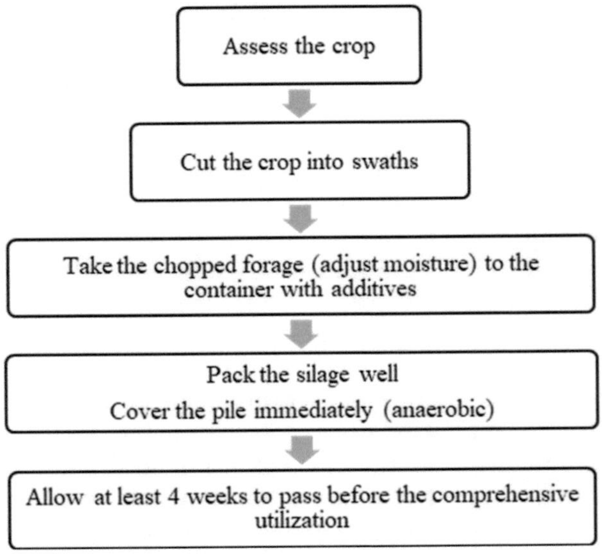

Figure 1.
The pictorial presentation of the steps involved in the ensilage formation.

Figure 2.
Large-scale storage in movable microbial ensilage membranes.

The goal of good fermentation is to maximize the production of lactic acid, thereby reducing the pH value and establishing an environment that is not suitable for the growth of harmful organisms. Good silage is made with less exposure to the air, allowing the *Lactobacillus* to start fermenting quickly. For example, modern biogas engineering adopts the following pretreatment devices to complete the anaerobic fermentation process (**Figure 2**).

According to the main biochemical and microbiological transformations occurring during the process, ensiling can be divided into three phases as follows.

1. Aerobic fermentation process

Anaerobic microbial pretreatment is a straw pretreatment technology utilizing microbial fermentation under anaerobic conditions. However, in the process of straw closure, there is more or less oxygen in the straw raw materials, which makes aerobic microorganisms grow and propagate in the first few days of fermentation. Through the activity of these aerobic microbes, a small amount of sugar and oxygen in the straw can be converted into carbon dioxide and water.

Finally, the oxygen becomes less and less until the oxygen content drops to zero. Aerobic microorganisms do not survive.

2. Enzymatic hydrolysis process

Due to the activity of microorganisms, various enzymes are produced, which destroy the structure of cellulose, hemicellulose, and lignin in the straw and make them degrade step by step to form various sugars. The enzymatic hydrolysis process of straw is relatively slow. With the increase of microbial reproduction and the improvement of microbial activity, the straw is gradually enzymatically hydrolyzed into sugars. Hemicellulose is most easily degraded throughout the enzymatic hydrolysis proofreading, while a larger amount of xylose, arabinose, glucose, mannose, and galactose is formed. When these sugars reach a certain concentration, microorganisms can use these sugars as substrates for acid production and fermentation.

3. Acidogenic fermentation process

Microorganisms utilize sugars in the straw as substrates and convert them into organic acids. After aerobic fermentation of the straw, oxygen is exhausted and aerobic microorganisms cannot survive. Under anaerobic conditions, anaerobic microorganisms cannot completely convert sugar substrates into water and carbon dioxide, but can only be decomposed into various organic acids, including hexanoic acid, propionic acid, lactic acid, and butyric acid. These organic acids are ionized in the straw to form a large amount of hydrogen ions, which acidify the straw and decrease the pH value. When the pH value drops to 4.5–5.0, acidity inhibits the activities of various microorganisms, thus slowing down the microbial activities and forming a good straw silage.

3.2 Additives

In addition to a small amount of LCB in raw materials, there are also a large number of spoilage bacteria, yeasts, and other flora that are not conducive to silage fermentation. Most of these floras are aerobic bacteria, while lactobacilli belong to anaerobic bacteria. In the early stage of fermentation, the LCB begin to propagate only after the oxygen in silage is consumed by these aerobic mold spoilage bacteria to form an anaerobic environment. In this process, silage will cause fever and serious occurrence of earthy or mildew taste, which are all adverse factors from silage. Artificial addition of additives can shorten the process, which makes LCB become the dominant bacteria group of silage and improve the overall quality of silage. At present, there are more than 200 silage additives used all over the world, which can be classified into four categories as a whole, namely, bacterial inoculants, enzyme, non-protein nitrogen, and preservatives.

3.2.1 Bacterial inoculants

LAB plays a major role in the fermentation process of silage, whose main strains include *Lactococcus* (*Streptococcus faecalis*, *Staphylococcus lactis*, *Lactobacillus lactis football*, etc.) and *Lactobacillus* (*Lactobacillus plantarum*, *Lactobacillus brucei*, *Lactobacillus casei*, *Lactobacillus brevis*, etc.). Inoculated LAB in silage can be divided into two types: homozygous and heterotypic fermentation. Homotypic fermentation mainly includes *Lactobacillus plantarum* and *Pediococcus lactis*, while heterotypic fermentation is mainly *Lactobacillus brucei* at present. The advantages of homozygous fermentation are that one glucose is converted into two lactic acids, which have strong acid-producing ability, fast pH value reduction, and less DM loss during fermentation. However, its disadvantage is that it produces a large amount of lactic acid, resulting in poor aerobic stability after cellar opening. After exposure to air, the silage pH value increases rapidly and the silage temperature increases accordingly. Heterotypic fermented convert 1 mol of glucose into 1 mol of lactic acid and 1 mol of acetic acid. Because the acidity of acetic acid is less than that of lactic acid, the speed of reducing pH value in fermentation is slower than that of the same type, and the loss during fermentation is higher than that of the same type of LCB. However, due to the production of a large amount of acetic acid, the aerobic stability after cellar opening is very good, up to 100 hours.

Lactobacillus inoculum, or LAB starter, is widely recognized and commercialized, mainly because it can reduce the loss of DM in the early stage of fermentation. However, during silage opening, this microbial inoculum failed in preventing thermal deterioration of silage and did not improve the aerobic stability of silage. What bother producers is certain yeasts that have a strong utilization of lactic acid, which are insensitive to high concentrations of lactic acid, are highly tolerant to low pH environments, and cannot inhibit its activity in acidic environments. They grow well in the acidic environment of silage. When the silage is opened, they are twice as active after exposure to air as in the sealed environment of silage. During the period of aerobic exposure, these yeasts can utilize lactic acid as an energy substance. The aerobic activity of yeasts will lead to the increase of silage pH due to the utilization and consumption of lactic acid, which weakens the acidic environment. In such an environment, other aerobic bacteria and molds will also activate, while the degree of silage corruption will deepen. Ultimately, the activity of aerobic microorganisms can not only lead to the loss of sugar, starch, and protein but also produce a lot of heat because of the large number of microbial reproduction and activity.

Nowadays, widely used and recognized inoculants contain mixed microbial inoculants of both homotypic LAB (to reduce pH) and heterotypic *Lactobacillus brucelli* (to inhibit yeasts), which can ensure both successful fermentation during anaerobic fermentation of silage and reduction of fever, deterioration, and spoilage during open pit.

3.2.2 Enzyme

Additional enzymes mainly refer to a variety of cell wall degrading enzymes, such as cellulase, hemicellulase, amylase, and pectinase. The purpose of adding enzymes is to reduce the fiber content in silage. In addition, more sugars can be degraded by enzymes for LAB fermentation. As enzymes contribute to the degradation of acidic and neutral washing fibers, lactic acid fermentation, dry matter recovery, and storage life are improved. Because of the large demand and high cost of enzyme preparation in the process of adding, it is seldom used in the actual production process.

3.2.3 Antiseptic additives

In order to reduce the pH value quickly, soften the silage, and facilitate the digestion of materials, dilute sulfuric acid or hydrochloric acid can be added to the silage. Preservatives are added to make the silage sink quickly, compact easily, increase the storage capacity, and make the silage crops stop breathing (biological oxidation) quickly, thus improving the success rate. Formic acid and propionic acid are commonly used. Benzoic acid and its sodium salt also have good bacteriostatic effect on molds in silage with dosage not exceeding 0.1%. In the United States and other places, calcium formate plus sodium sulfite was used for silage. You can also choose some other antifungal agents, such as sorbic acid and its potassium salt, as appropriate, but the price is higher. When using mold inhibitors, spray them evenly on the shredded silage raw materials as far as possible, compact them in layers, and seal them in good condition.

3.3 Quality evaluation of the silage

The standardization and development of silage quality grading may help to accurately assess the quality, which is one of the key measures to improve the quality of forage. Generally speaking, quality evaluation of silage includes sensory evaluation and chemical analysis [25]. Sensory evaluation is simple and feasible, but subjective; laboratory evaluation of chemical analysis involves many indicators, which can be quantitatively compared objectivity. According to the color, odor, texture, structure, and other indicators of silage, the quality of silage was evaluated by sensory operation. Chemical analysis evaluation is to quantitatively evaluate the chemical composition of silage through instrumental analysis, while the main indicators are pH value, organic acids (lactic acid, acetic acid, butyric acid, etc.), ammonia nitrogen, total nitrogen, alcohols, and other fermentation parameters.

3.3.1 Sensory evaluation

Using the sensory evaluation method (**Table 1**), we can quickly and intuitively judge whether the silage material is deteriorating.

The author has carried out anaerobic pretreatment on dry corn stover (**Figure 3**). It can be seen that after 28 days of storage, there is no mildew in the samples of dry corn stover silage. The color of the silage material is yellow and has a strong acid flavor, indicating that the process was successful.

3.3.2 Laboratory measurements of the silage

1. **Lactic acid**: the goal of good fermentation is to maximize the production of lactic acid.

   Lactic acid is the strongest fermentation acid in the fermentation process, which has the best effect in reducing pH value. Rapid reduction of pH value helps to reduce protein decomposition, increase the acidolysis of hemicellulose, and reduce other useless microbial activities. The high ratio of lactic acid and lactic acid/acetic acid indicates that good fermentation has taken place.

Sensory evaluation	Quality		
	High	Medium	low
Odor	Sour smell, aroma, and strong sour taste, like distillers' grains and pickles	Less acidity and fragrant but not strong flavor	Strong butyric acid or pungent odor, mildew odor
Color	Green or yellow-green, similar to the color of the raw material	Yellowish brown or dark green	Brown or black
Texture	Stems and leaves are well preserved, and the hands are loose, soft, and slightly moist	Soft but slightly dry or slightly watery, with separable stems and leaves	Stems and leaves rot, stick together, or are loose, dry, and hard

Table 1.
The sensory evaluation of silage.

Figure 3.
The surface morphology of the starting material and the resulting silage.

2. **Acetic acid**: the formation of acetic acid usually occurs in the first 2 to 3 days of silage.

When the pH value drops below 5.00, the *Lactobacillus* begins to grow and takes over the fermentation process. Therefore, in typical silage, the production of acetic acid is helpful to activate the production of lactic acid. In the process of opening cellar to take silage, the degradation of lactic acid by yeast can also produce acetic acid. The concentration of acetic acid is usually low (<3%). It has some antifungal properties, which help to prolong the retention time of the silage.

3. **Propionic acid**: usually found in only a small proportion with silage.

A high level of propionic acid with silage indicates a large error in some places. It is normal that propionic acid is higher when adding propionic acid additive.

4. **Butyric acid**: butyric acid is produced by *Clostridium perfringens*. If the silage is too wet (DM <30%), the bacteria will multiply.

In wet silage, the acid produced by *Lactobacillus* may not be enough to reduce the pH value to prevent the growth of *Clostridium*. *Clostridium* can also ferment lactic acid to produce butyric acid and decompose amino acid to produce excessive ammonia. Both of these effects lead to an increase in pH and further deterioration of silage.

The results of the author's study are as follows: The pH value (4.22), pretreatment time (4 weeks), and the content of lactic acid in dry corn stover silage (4.32%) should be considered as important indicators of the success of microbial ensiling [26].

4. Bioconversion of silage to biogas

In exploring the effect of pretreatment with wastes, many studies have explored the structure and hydrolysis degree of pretreated materials, which as an important index to evaluate the advantages and disadvantages of pretreatment [27]. Like physical and chemical pretreatment, biological pretreatment can also loosen the structure of lignocellulose (**Figure 4**), but it cannot be ignored that the change of lignocellulose structure is only a part of the indicators to evaluate the advantages and disadvantages. It is also an important part of the transformation ability of substrates and the type of metabolites. Various organic acids, especially lactic acid, produced by microbial metabolism during anaerobic pretreatment are important intermediates in

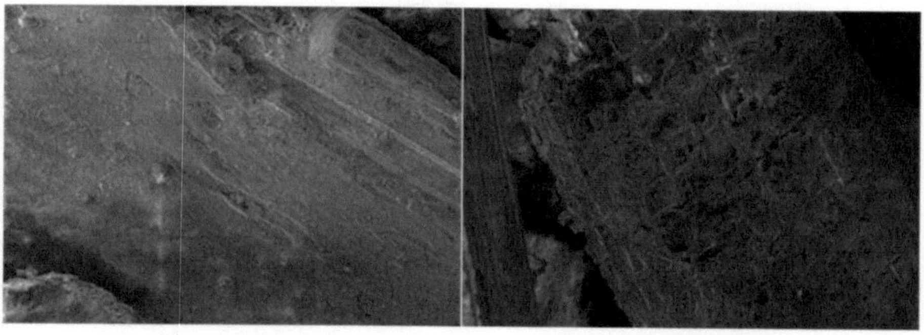

Figure 4.
Scanning electron micrographs of non-pretreated and pretreated dry corn stover.

the AD process [28], which can improve the biogas yield and accelerate the process of biogas fermentation [29]. In addition, the organic acids can effectively neutralize the ammonia and other basic substances accumulated during AD to ensure that the pH value of the reaction system is maintained in a stable range [30]. This indicates that the bio-pretreatment method which can metabolize organic acids can effectively improve the biogas yield. However, not all biological pretreatments can convert LCB into organic acids in straw, or only have low conversion capacity, which may limit the significant improvement of subsequent gas production [31].

In the process of AD, raw materials are not only digestive substrates but also the source of nutrients for the survival of anaerobic microorganisms. The character of raw materials determines the time and the biogas yield of AD [32]. AD is a coordinated and mutually restricted metabolic process among starch, protein, fat, and lignocellulose. The types and quantities of raw material organic matter play a decisive role in the degradation process, raw material utilization efficiency, and biogas yield of AD. For example, AD with single LCB has the problem of imbalance between carbon and nitrogen. The C/N ratio of straw stalk is close to 70:1, whereas the best C/N ratio for AD is 20–30:1 [33]. Due to the low C/N ratio, ammonia nitrogen inhibition occurs in AD with single pig manure, which makes it difficult to form the optimal growth state required by biogas-producing microbiota [34]. Hydrolysis pretreatment is a limiting step, because the structure of the cell wall of the residual sludge can inhibit the hydrolysis of intracellular degradable substances. Gas production from conventional sludge digestion often requires a longer residence time [35].

To solve the above problems, mixed pretreatment with difficult and easy to decompose organic matter is one of the hotspots in the field of AD in recent years. Mixed AD can not only effectively regulate the nutritional balance of single raw materials but also improve the bioconversion rate of materials [36]. The author carried out anaerobic microbial ensilage pretreatment using dry corn stover mixed with pig manure and residual sludge, respectively, and then investigated the biogas yield on this basis. The gas production rate from total solids (TS) of untreated dry straw was only 296 mL/g, and the gas production index has been greatly improved through mixed microbial ensilage. The effect of the dry corn stover and pig manure microbial ensilage pretreatment was the best, whereby the gas production rate of TS reached 599 mL/g, and the average volumetric gas production rate was 0.86 L/(L·d). The AD of corn stover, excess sludge, and pig manure can be used to alleviate the nitrogen limitation when using silage as the main raw material. Broad bioconversion of such raw materials will play a decisive role in solving the problems of burning straw, residual sludge landfill, and non-point source pollution from livestock and poultry manure.

The combined AD of different silage materials has the following points:

1. adjusting the organic nutrients in the fermentation substrate;

2. producing organic acids and intermediate metabolites such as ammonia nitrogen can undergo partial neutralization reaction to maintain the stability of pH value in the process and promote the smooth progress of AD; and

3. increasing the degradation rate of raw materials to shorten the whole fermentation time.

5. Conclusions

Solid-state anaerobic microbial ensilage pretreatment is one of the effective means to improve the biogas yield of raw materials. Other pretreatment methods only change the composition and structure. Biological pretreatment can not only change the composition and structure but also have the ability to ferment and metabolize decomposed substrates into other small molecules, and change the composition and structure. It takes a certain time for fermentation to metabolize small molecule nutrients; so, exploring the optimal pretreatment time can improve the efficiency. Because of the particularity of biological pretreatment, it is also of great significance to explore its impact on the structure, activity, and functional expression of microbial community in the subsequent AD. Understanding the impact of biological pretreatment from multiple angles can better optimize the biological pretreatment conditions, thereby improving the efficiency and enhancing the gas production efficiency for biogas engineering.

The high efficiency of physical pretreatment is accompanied by high energy consumption, and chemical pretreatment is also facing environmental pressure. Biological pretreatment also needs to constantly improve its pretreatment efficiency. Therefore, more and more studies began to focus on mixed pretreatment, that is, to absorb the advantages of different pretreatments to make up for the shortcomings, to achieve a higher hydrolysis rate while reducing energy consumption and reaction time, and to reduce pretreatment costs and environmental hazards. For mixing pretreatment with different materials, screening suitable microbial communities will be a future study direction to ensure a controllable and stable treatment process.

Acknowledgements

The author is grateful to Zhengzhou Key Laboratory of Metabolic Engineering and Systems Biology for providing experimental conditions.

This work was supported by the Doctoral Scientific Research Foundation of Zhengzhou University of Light Industry (No. 0123/13501050066) and Key Scientific Research Project Plan in colleges and universities of Henan Province (No. 20B416003).

Conflict of interest

The authors declare no conflict of interest.

Author details

Xu Yang
School of Food and Bioengineering, Zhengzhou University of Light Industry, Zhengzhou, China

*Address all correspondence to: yangxu@zzuli.edu.cn

References

[1] Tian SQ, Zhao RY, Chen ZC. Review of the pretreatment and bioconversion of lignocellulosic biomass from wheat straw materials. Renewable and Sustainable Energy Reviews. 2018;**91**:483-489. DOI: 10.1016/j.rser.2018.03.113

[2] Yang L, Ge X, Wan C, Yu F, Li Y. Progress and perspectives in converting biogas to transportation fuels. Renewable and Sustainable Energy Reviews. 2014;**40**:1133-1152. DOI: 10.1016/j.rser.2014.08.008

[3] Nkoa R. Agricultural benefits and environmental risks of soil fertilization with anaerobic digestates: A review. Agronomy for Sustainable Development. 2014;**34**:473-492. DOI: 10.1007/s13593-013-0196-z

[4] Zheng Y, Zhao J, Xu F, Li Y. Pretreatment of lignocellulosic biomass for enhanced biogas production. Progress in Energy and Combustion Science. 2014;**42**(1):35-53. DOI: 10.1016/j.pecs.2014.01.001

[5] Ai P, Zhang XZ, Ran Y, Meng L, Elsayed M, Fan QZ, et al. Biomass briquetting reduces the energy loss during long-term ensiling and enhances anaerobic digestion: A case study on rice straw. Bioresource Technology. 2019;**292**:121912. DOI: 10.1016/j.biortech.2019.121912

[6] Matheri AN, Ntuli F, Ngila JC, Seodigeng T, Zvinowanda C, Njenga CK. Quantitative characterization of carbonaceous and lignocellulosic biomass for anaerobic digestion. Renewable and Sustainable Energy Reviews. 2018;**92**:9-16. DOI: 10.1016/j.rser.2018.04.070

[7] Mood SH, Golfeshan AH, Tabatabaei M, Jouzani GS, Najafi GH, Gholami M, et al. Lignocellulosic biomass to bioethanol, a comprehensive review with a focus on pretreatment. Renewable and Sustainable Energy Reviews. 2013;**27**(6):77-93. DOI: 10.1016/j.rser.2013.06.033

[8] Peng J, El-FatahAbomohra A, Elsayed M, Zhang X, Fan Q, Ai P. Compositional changes of rice straw fibers after pretreatment with diluted acetic acid: Towards enhanced biomethane production. Journal of Cleaner Production. 2019;**230**(1):775-782. DOI: 10.1016/j.jclepro.2019.05.155

[9] Sun SL, Wen JL, Ma MG, Sun RC. Enhanced enzymatic digestibility of bamboo by a combined system of multiple steam explosion and alkaline treatments. Applied Energy. 2014;**136**(C):519-526. DOI: 10.1016/j.apenergy.2014.09.068

[10] Elsayed M, Abomohra EF, Ai P, Jin K, Fan Q, Zhang Y. Acetogenesis and methanogenesis liquid digestates for pretreatment of rice straw: A holistic approach for efficient biomethane production and nutrient recycling. Energy Conversion and Management. 2019;**195**(September):447-456. DOI: 10.1016/j.enconman.2019.05.011

[11] Chen W, Tu Y, Sheen HK. Disruption of sugarcane bagasse lignocellulosic structure by means of dilute sulfuric acid pretreatment with microwave assisted heating. Applied Energy. 2011;**88**(8):2726-2734. DOI: 10.1016/j.apenergy.2011.02.027

[12] Lima CSS, Conceio MM, Silva FLH, Lima EE. Characterization of acid hydrolysis of sisal. Applied Energy. 2013;**102**(2):254-259. DOI: 10.1016/j.apenergy.2012.09.061

[13] Elsayed M, Abomohra AE, Ai P, Wang D, El-Mashad HM, Zhang Y. Biorefining of rice straw by sequential fermentation and anaerobic digestion for bioethanol and/or biomethane

production: Comparison of structural properties and energy output. Bioresource Technology. 2018;**268**:183-189. DOI: 10.1016/j.biortech.2018.07.130

[14] Sun H, Cui X, Stinner W, Shah GM, Cheng HC, Shan SD, et al. Synergetic effect of combined ensiling of freshly harvested and excessively wilted maize stover for efficient biogas production. Bioresource Technology. 2019;**285**:121338. DOI: 10.1016/j.biortech.2019.121338

[15] Dixon RA. Microbiology: Break down the walls. Nature. 2013;**493**(7430):36-37. DOI: 10.1038/493036a

[16] Wan C, Li Y. Fungal pretreatment of lignocellulosic biomass. Biotechnology Advances. 2012;**30**(6):1447-1457. DOI: 10.1016/j.biotechadv.2012.03.003

[17] Ghosh A, Bhattacharyya BC. Biomethanation of white rotted and brown rotted rice straw. Bioprocess and Biosystems Engineering. 1999;**20**(4):297-302. DOI: 10.1007/s004490050594

[18] Taniguchi M, Suzuki H, Watanabe D, Sakai K, Hoshino K, Tanaka T. Evaluation of pretreatment with *Pleurotus ostreatus* for enzymatic hydrolysis of rice straw. Journal of Bioscience and Bioengineering. 2005;**100**:637-643. DOI: 10.1263/jbb.100.637

[19] Zhong WZ, Zhang ZZ, Qiao W, Fu P, Liu M. Comparison of chemical and biological pretreatment of corn straw for biogas production by anaerobic digestion. Renewable Energy. 2010;**36**:1875-1879. DOI: 10.1016/j.renene.2010.12.020

[20] Mackulak T, Prousek J, Svorc L, Drtil M. Increase of biogas production from pretreated hay and leaves using wood-rotting fungi. Chemical Papers. 2012;**66**:649-653. DOI: 10.2478/s11696-012-0171-1

[21] Herrmann C, Heiermann M, Idler C. Effects of ensiling, silage additives and storage period on methane formation of biogas crops. Bioresource Technology. 2011;**102**(8):5153-5161. DOI: 10.1016/j.biortech.2011.01.012

[22] Peter W. Biogas production: Current state and perspectives. Applied Microbiology and Biotechnology. 2010;**85**(4):849-860. DOI: 10.1007/s00253-009-2246-7

[23] Cui ZF, Shi J, Wan CX, Li YB. Comparison of alkaline- and fungi-assisted wet-storage of corn stover. Bioresource Technology. 2012;**109**(4):98-104. DOI: 10.1016/j.biortech.2012.01.037

[24] Piotr OP, Anne BT, Jens ES. Ensiling – Wet-storage method for lignocellulosic biomass for bioethanol production. Biomass and Bioenergy. 2011;**35**(5):2087-2092. DOI: 10.1016/j.biombioe.2011.02.003

[25] Daems F, Decruyenaere V, Agneessens R, Lognay G, Romnee JM, Froidmont E. Changes in the isoflavone concentration in red clover (*Trifolium pratense* L.) during ensiling and storage in laboratory-scale silos. Animal Feed Science and Technology. 2016;**217**:36-44. DOI: 10.1016/j.anifeedsci.2016.04.008

[26] Yang X, Chang C, Cheng JY, Li HL, Ma XJ. The integrated process of microbial ensiling and hot-washing pretreatment of dry corn stover for ethanol production. Waste and Biomass Valorization. 2018;**9**:2031-2040. DOI: 10.1007/s12649-017-0007-x

[27] Li HQ, Qu YS, Yang YQ, Chang SL, Xu J. Microwave irradiation – A green and efficient way to pretreat biomass. Bioresource Technology. 2016;**199**:34-41. DOI: 10.1016/j.biortech.2015.08.099

[28] Bremond U, de Buyer R, Steyer JP, Bernet N, Carrere H. Biological pretreatments of biomass

for improving biogas production: An overview from lab scale to full-scale. Renewable and Sustainable Energy Reviews. 2018;**90**:583-604. DOI: 10.1016/j.rser.2018.03.103

[29] Franco RT, Buffière P, Bayard R. Ensiling for biogas production: Critical parameters. A review. Biomass and Bioenergy. 2016;**94**:94-104. DOI: 10.1016/j.biombioe.2016.08.014

[30] Annukka P, Pekka M, Seija J, Stoddard FL, Maritta K, Liisa V. Evaluation of preservation methods for improving biogas production and enzymatic conversion yields of annual crops. Biotechnology for Biofuels. 2011;**4**(1):20-32. DOI: 10.1186/1754-6834-4-20

[31] Hua BB, Dai JL, Liu B, Zhang H, Yuan XF, Wang XF, et al. Pretreatment of non-sterile, rotted silage maize straw by the microbial community MC1 increases biogas production. Bioresource Technology. 2016;**216**:699-705. DOI: 10.1016/j.biortech.2016.06.001

[32] Yang X, Chang C, Li HL, Ma XJ. Mixed ensiling pretreatment and semi-continuous fermentation of dry corn stover and wastes for biogas production. Journal of Chemical Engineering of Chinese Universities. 2017;**31**(4):899-905. DOI: 10.3969/j.issn.1003-9015.2017.04.020

[33] Paudel SR, Banjara SP, Choi OK, Park KY, Kim YM, Lee JW. Pretreatment of agricultural biomass for anaerobic digestion: Current state and challenges. Bioresource Technology. 2017;**245** (Part A):1194-1205. DOI: 10.1016/j.biortech.2017.08.182

[34] Wu SX, Yao WY, Zhu J, Miller C. Biogas and CH_4 productivity by co-digestion swine manure with three crop residues as an external carbon source. Bioresource Technology. 2010;**101**(11):4042-4047. DOI: 10.1016/j.biortech.2010.01.052

[35] Kim HW, Nam JY, Shin HS. A comparison study on the high-rate co-digestion of sewage sludge and food waste using a temperature-phased anaerobic sequencing batch reactor system. Bioresource Technology. 2011;**102**(15):7272-7279. DOI: 10.1016/j.biortech.2011.04.088

[36] Tsapekos P, Kougias PG, Frisona A, Ragab R, Angelidakia I. Improving methane production from digested manure biofibers by mechanical and thermal alkaline pretreatment. Bioresource Technology. 2016;**216**:545-552. DOI: 10.1016/j.biortech.2016.05.117

Chapter 5

Modelling of Carbon Monoxide and Carbon Dioxide Methanation under Industrial Condition

Artur Wodołażski

Abstract

The development of methanation technology is supported by detailed modeling and process simulation to optimize the design and study of its reaction dynamic properties. The chapter presents a discussion of selected catalysts and its kinetic models in the methanation reaction. The development models of fixed-bed reactors in the methane synthesis were also presented. Chemical and physical modeling of methanation reactions with optimization, exploitation, and the analysis of critical processes in time is an important contribution to the technology modernization.

Keywords: methanation, process design, reactor modeling, catalytic methanation

1. Introduction

Effective transformation of carbon dioxide, which arises when burning fuels in industrial installations on substitution natural gas (SNG), is one of the challenges of the twenty-first century implemented in new research and implementation projects, especially that the pressure of the European Union (EU) to reduce CO_2 emissions increases with every decade. In case of facing problems with carbon capture and storage (CCS) technologies, which are expensive, and at the same time due to the lack of social acceptance for the transport of compressed CO_2 and its underground storage, the European Commission has recently started to suggest the implementation of carbon capture and utilization (CCU) technology or transformation of carbon dioxide into socially acceptable products [1–3].

1970s and 1980s are years, which we were looking for a solution that would turn sulfur into a product like gypsum. Now, we face a similar problem of what to do with carbon dioxide. Industrial methanation processes, which are based on the transformation of carbon dioxide, e.g., from coal-fired power plants and chemical and metallurgical plants or from cement plants to methane as a result of hydrogen-derived water electrolysis fed with surplus cheap energy or from coal gasification [4].

The purpose of the use of surplus energy from wind farms for hydrogen production is to address the problems related to renewable energy sources. A serious drawback of such installations is the lack of human influence on the current volume of production, which depends directly on the forces of nature, i.e., in the case of a wind farm depends on the instantaneous strength of the wind. Hydrogen produced in the process of electrolysis of water, the additional product of which is oxygen (it can be sold or, for example, used in power coal gasification or for oxygen combustion—in a method that simplifies CO_2 capture). Although, carbon dioxide will come from pilot installations for capturing CO_2 from exhaust gases. CO_2 processing as part of previous research projects in their power plants (mobile installation of amine CO_2 capture in Łaziska Power Plant in a project shared with Tauron PE and variable pressure installation in Łagisza Power Plant) is one of the examples of industrial methanation [5].

A research and development organization CEA deals with CO_2 methanation in modular structural reactors. The first test reactor was started with a capacity of 3–4 m^3 CO_2/h at the inlet. One of the companies dealing in the production, purchase of equipment, and their integration with the entire installation will be prepared by West Technology & Trading (deals in the design and construction of industrial installations, among others) [6].

Industrial research and process optimization play an important role in the development and implementation of industrial installations converting CO_2 to methane, which enables effective conversion of CO_2 into methane. CO_2 management options using surplus energy from renewable sources, as well as the possibility of energy storage by converting surplus electricity into a natural gas substitute. All this also increases the stability of the entire energy system, and above all, it is socially acceptable [7].

Among the advantages of the proposed solution, project participants also mention the ease of the use of captured CO_2 without its high compression, transport by pipeline or underground storage, as well as reduction of dependence on natural gas and its imports from outside Europe through the use of gas networks. The risks associated with the project are related to technological issues. This is, for example, the problem of long-term stabilization of the system's efficiency (e.g., catalyst aging) and its durability, as well as the necessity of successful scaling from a demonstration to a commercial installation, for example, on a scale corresponding to the capacity of converting CO_2 from a power unit. However, the key aspect in technology is the type of used catalysts [8]. The chapter presents a model of fixed-bed methanation reactors for one-, two-, and three-dimensional models with a pseudohomogeneous or heterogeneous phase with the application of various kinetic models. The main kinetic models include Kopyscinski, Xu-Froment, van Herwijnen, and van Doesburg. When creating mathematical models, various types of software were used, including MATLAB, OpenFOAM, FEMLAB, Fortran, Modelica, as well as Aspen Plus and ACM. Modeling the methanation reaction allows to effectively predict the rate of methanation reaction depending on the temperature, pressure, or initial concentrations of reagents. Therefore, model research focuses on the modeling of reactions and reactors; however, the reaction mechanism or the degree of catalyst deactivation factors significantly influences the development of the entire process.

2. Methanation of carbon monoxide catalysis and process kinetics

Carrier metal catalysts play a large role in modern catalysis. They strive to ensure high activity and selectivity in the catalyzed reaction while maintaining stability to ensure their long-term use. The activity of the catalysts is mainly related to the degree of dispersion of the active phase on the support and the state and nature of the interactions between the catalysts and the reactants. Appropriate selection of the catalytic system, preparation conditions, and reaction conditions determines the stability of the catalysts and thus protects against catalyst sintering or poisoning [9].

The CO_2 methanation reaction seems to be interesting for a number of reasons. It allows, among others, the use of waste hydrogen for methane synthesis and removal of traces of CO and CO_2, which are strong poisons of iron catalysts in the ammonia synthesis.

The researchers' interest is usually to determine the effect of various carriers and metals on the efficiency and selectivity of catalysts in the CO_2 methanation reaction, with particular reference to the catalysts obtained on the basis of ruthenium, as well as the selection of optimal conditions for this process [10]. Much work has been devoted to the study of the mechanism of methanation reactions of both oxide and carbon dioxide [11–15]. Carbon oxides are a strong poison of ammonia synthesis catalysts. The methanation process allows to reduce the value of the sum of carbon oxides in the synthesis mixture from 0.2 to 0.5% vol. to below 2 ppmv.

The course of methanation of carbon oxides is described by the following equations of the reaction:

$$CO + 3H_2 \leftrightarrow CH_4 + H_2O \qquad \Delta H_{298} = -206,2kJ/mol \tag{1}$$

$$CO_2 + 4H_2 \leftrightarrow CH_4 + 2H_2O \qquad \Delta H_{298} = -165,0\ kJ/mol \tag{2}$$

The equilibrium state of both reactions shifts very far to the right side of the equations, and, e.g., at 310°C under atmospheric pressure, the equilibrium concentrations of oxides are about 104 ppm.

For the kinetics and mechanism of CO and/or CO_2 methanation, a cycle of works was devoted. The processes were investigated at atmospheric pressure in a glass cell-free reactor on a NKM-4A catalyst crushed to a grain size of 0.5–1 mm. It was found that CO methanation was much faster than CO_2 methanation according to a different kinetic relationship. The reaction kinetics in the carbon monoxide mixture were complex. CO methane was very poorly controlled by the presence of CO_2, whereas CO_2 methane was strongly inhibited by the presence of CO. From the work, it was concluded that CO methanation-specific activity of nickel depends not only on the nature of the catalyst carrier, but on the methanation of CO_2, and for various catalysts, it varies up to tenfold. Literature information may be an insufficient source of knowledge for industrial process designers as well as for catalyst manufacturers. Therefore, modeling works and reliable information on process kinetics collected on the basis of real reaction rate measurements are important [16].

In most industrial methanation installations, nickel deposited on alumina is used as a catalyst. As a consequence, CO_X methanation reactions on nickel are well understood and described. The industrial process is carried out in the temperature range 250–320°C, under pressure from 1 atm to even several hundreds of atm. The development of more active systems in the methanation reaction may allow for the precise purification of gas from traces of CO_X using a smaller amount of catalyst or reduce the cost of the catalytic charge. This has a major impact on reducing the operating cost of the installation. It is believed that ruthenium is a more active nickel catalyst for the methanation of carbon oxides. In contrast to nickel, which undergoes deactivation at low temperatures due to the formation of carbonyls, ruthenium is stable over a wide temperature range. Commercial ruthenium catalysts for low-temperature methanation of carbon oxides are offered by such producers as Alvigo (RKM-3) or Süd-Chemie (METH 150). Ruthenium systems based on various supports were investigated: aluminum oxide, titanium oxide, or silica. Promising results were obtained using activated carbon as active carriers.

Catalyst (wt.%)	T (°C)	P (bar)	Kinetic approach	Author
Ni/SiO_2 (60:40%)	260–400	1	Langmuir-Hinshelwood	Burger and Koschany [18]
Ni/SiO_2 (60:40%)	280–400	30	Langmuir-Hinshelwood	van Meerten [19]
Ni/Cr_2O_3 (62:38%)	160–180	1	Langmuir-Hinshelwood	Kopyscinski [20]
Ni/Al_2O_3 (28:72%)	200–230	1	Power law	Chen [21]
Ni/Al_2O_3 (9% Ni, 1.5% Pt)	260–300	1–5	Power law	Hayes [22]
Ni/SiO_2 (58:42%)	275–320	17	Langmuir-Hinshelwood	Karolyi [23]
$Ni/MgAl_2O_3$ (50:50%)	300–400	10	Langmuir-Hinshelwood	Zhang [24]

Table 1.
Kinetic models used in methanation reaction.

Nickel catalysts, supported on a carrier, are commonly used in many important industrial hydrogenation and hydrogenolysis processes, such as methane steam reforming or methanation of synthesis gas, due to the ease of their preparation, high activity, and economic aspects [17]. Nickel catalysts are preferred in the hydrogenation of vegetable oils, edible oils, and the fat industry. Du Bois and other colleagues have designed a nickel complex that is surrounded by amine ligands, which, like in the case of enzymes, enable efficient transfer of protons to nickel atoms. This catalyst is stable and very efficient. Kinetic models used in methanation reaction are presented in **Table 1**.

Catalytic processes, involving a fixed catalyst, so-called contact processes, are the basis of the chemical industry, especially the large-volume industry. It is possible to mention the synthesis of ammonia, oxidation of SO_2 to SO_3, steam reforming of hydrocarbons, conversion of carbon monoxide, catalytic cracking, hydrocracking, or hydrodesulfurization. The catalytic methanation of carbon oxides is used on a large scale for the purification of hydrogen in ammonia synthesis plants. Recently, it has also been proposed to use this reaction for the purification of hydrogen fuels for low-temperature fuel cells.

Kinetics of carbon monoxide methanation on nickel catalysts was examined by Sehested [25], Hayes [22], and van Meerten [19]. Sehested et al. [25] investigated the methanation activity and adsorption of CO and H_2 on a supported Ni catalyst and micrometer-sized Ni threads with a mass of up to 0.4 g at low CO partial pressures (pCO < 25 mbar) in single-pass and recirculation reactors at a total pressure of 1.4 bar (mostly H_2) and temperatures of $225 < T < 400°C$. Fujita et al. [26] calculated CO and C coverages in transient response experiments during CO and CO_2 methanation at atmospheric pressure, 200°C, and varying H_2/CO_X ratios in a differential flow reactor containing 1.0 g of Ni catalyst. van Meerten et al. [19] studied kinetics and the mechanisms of methanation of CO on a 5% Ni/SiO_2 catalyst in a differential flow microreactor at $187 < T < 567°C$, $133 < pCO < 8.7 \times 10^4$. Hayes et al. [22] used a tubular microreactor with a sample mass of up to 1 g at a pressure of 1.3 bar and temperatures of $270 < T < 380°C$ for dynamic response studies of alumina-supported Ni catalysts prepared by impregnation and coprecipitation.

3. Models of methanation process: fixed-bed reactor development

Chen and Yu [21] built a numerical model of a fixed-bed reactor in which they studied the reaction of CO_2 methanation for the production of synthetic natural gas (SNG).

They developed a numerical model to study the reaction of CO_2 methanation in a tubular reactor with a fixed bed in which the process efficiency was measured. A catalyst consisting of Ni and Ru was used. The reagent temperature in this test was used as the primary parameter at the reactor inlet. Based on the obtained results, it was found that the optimal temperature of the reactant at the inlet appeared at the maximum CO_2 concentration. When the inlet temperature rises above the optimum value, the CO_2 conversion decreases due to the inverse of the Sabatier reaction. The CO_2 conversion can be increased by increasing the working pressure, reactor size, and H_2/CO_2 ratio. CO_2 conversion may also be increased by reducing the feed rate of the reactant. The best H_2 yield can be obtained using a stoichiometric amount of H_2. The introduction of an inert gas into the reactant reduces the CO_2 conversion due to the decrease of CO_2 and H_2 partial pressure. With an isolated reactor and insufficient heat dissipation and high temperature in the reactor, the Sabatier reaction is reversed. The Sabatier reverse reaction caused a decrease in the CO_2 conversion. At low inlet temperature, the heat can be removed, which can lower the reaction temperature, leading to low CO_2 conversion. This confirms the thesis that the Ru-based catalyst has a better performance than the nickel-based catalyst. Bai and Wang [24] studied the effects of the bed aspect ratio, inlet feed temperature, and the pressure in the methanation reactor using coal to synthetic natural gas (SNG) as a model process. A two-dimensional pseudo-homogeneous model was established to simulate the fixed-bed methanation reactor. The numerical model was solved by MATLAB code and validated by an industrial sidestream test. Results show that a bed aspect ratio of 2–3 is helpful to reduce the heat loss below 700°C. A feed temperature of 400°C can accelerate reaction rate and lead to a higher bed temperature of ~800°C. The CO hydrogenation process is dated by operating pressure. The location of the hotspots is equal to 16,000 h^{-1}. When the hydrogen-to-carbon ratio (defined as $H_2/(3CO + 4CO_2)$) in the feed increases to 2, the CO conversion can be promoted. An increase of the steam-to-gas ratio from 0.19 to 0.4 can effectively control the adiabatic temperature with a hotspot temperature of ~650°C.

3.1 One-dimensional pseudohomogeneous fixed-bed reactor model

Dissinger et al. [27] built a one-dimensional pseudohomogeneous reactor model with adiabatic temperature characteristics, where he used the Fortran programming language to solve numerical equations. van Doesburg [28] built a 1-D pseudohomogenic model using the van Herwijnen kinetic equation with adiabatic temperature characteristics. Er-rbib et al. [29] derived the 1-D model of a pseudohomogenic isothermal reactor using the Kopyscinski kinetic equation [20].

The one-dimensional (1-D) pseudo-homogeneous model of a tubular reactor is described by a system of equations presenting mass and heat balance according to equations:

$$\text{Mass balance}: \quad \frac{\partial(uc_i)}{\partial x} = \rho_{bed} \sum_{j=1}^{i} v_{i,j} r_j \tag{3}$$

$$\text{Energy balance}: \quad uc_t c_p \frac{\partial T}{\partial x} = \rho_{bed} \sum_{j=1}^{i} r_j(-\Delta_R H_j) - \frac{4}{d} U(T - T_C) \tag{4}$$

where U is the effective overall heat transfer coefficient, W/m^2 K; c_p is the specific heat capacity, J/K; T_C is the cooling temperature, K; and d is the tube diameter, m.

This model has gotten the following assumptions: pressure drop along the axial reactor coordinate is neglected, and the partial pressures of the components are calculated assuming ideal gas behavior. The specific heat capacity Cp_i is calculated by the Shomate equation.

The 1-D model describing the methanation process, taking into account the mass and heat equations, was formulated by Wasch and Froment [30]. It assumed a series of thermal resistances regarding the heat transfer coefficient and radial heat dissipation. According to the model, radial heat dispersion and effective heat transfer coefficient are characterized by static and dynamics depending on the heat flow areas. When modeling the methanation reaction, the model must meet the following assumptions:

1. Gas mixture is considered ideal.
2. The system is in a steady-state condition.
3. Also, the cross section of the reactor is ideal.
4. Axial mass and heat transfer are assumed to be negligible.

Khorsand and Marvast [31] modelling a methanation catalytic reactor in ammonia unit. The catalytic reactor is similar to that of the steam reforming one. They used orthogonal collocation method to modeling catalytic methanation reaction.

3.2 Two-dimensional pseudohomogeneous fixed-bed reactor model

Schlereth and Hinrichsen [32] used the MATLAB software to build a two-dimensional reactor model using Xu and Froment kinetic equation with a polytropic thermal characteristic.

The two-dimensional (2-D) pseudohomogeneous model of a tubular reactor is described by a system of equations presenting mass and heat balance in accordance with the equations:

$$\text{Mass balance}: \quad \frac{\partial(uc_i)}{\partial x} = \rho_{bed} \sum_{j=1}^{i} v_{i,j} r_j + D_{eff}\left(\frac{\partial^2 c_i}{\partial r^2} + \frac{1}{r}\frac{\partial c_i}{\partial r}\right) \tag{5}$$

$$\text{Energy balance}: \quad u c_t c_p \frac{\partial T}{\partial x} = \rho_{bed} \sum_{j=1}^{i} r_j\left(-\Delta_R H_j\right) + \Lambda_{eff}\left(\frac{\partial^2 T}{\partial r^2} + \frac{1}{r}\frac{\partial T}{\partial r}\right) \tag{6}$$

They consider the tube's radial coordinate next to the axial one and generally assume axial symmetry, that is, variations of packed beds in angular directions are excluded from consideration. The two-dimensional model gives more detailed information for a quantitative evaluation and the comparison with experimental results. Heat removal from the reactor becomes an important issue to prevent the thermal runaway problem. The model assumes constant values of the property parameter describing pseudo-homogeneous continuity, i.e., the porosity and dispersion coefficients are independent of the radial position of the model. Therefore, it is characterized by heat transfer from the coolant to the inner wall of the pipe (the resistance caused by conduction in the metal wall is neglected). In this model, the introduction of a heat transfer coefficient through the artificial wall for the interior of the pipe can be avoided because the heat dissipation factor is a function of the radial position and therefore also captures the resistance adjacent to the pipe wall.

The wall heat transfer coefficient and effective heat dispersion coefficient are identical to the contributions of the effective heat transfer coefficient for the 1-D model. The molecular Péclet number is used, and the diffusion coefficient is substituted for the thermal conductivity.

3.3 Heterogeneous reactor model

The 1-D model of the heterogeneous reactor was derived by Bader [33] with adiabatic temperature characteristics using Modelica Fluid software.

Heterogeneous reactors can be simplified by pseudo-homogeneous reactor models, if the effect of external mass transfer and pore diffusion limitations are negligible.

Khorsand et al. [31] used the MATLAB software, where using the kinetic model Xu and Froment to build a 1-D heterogeneous model with adiabatic temperature framework where the catalyst was maintained in isothermal conditions. While in pseudo-homogeneous models, it assumes spontaneously that neither concentration.

There are also no temperature differences between the gas phase and the solid catalyst granules, because only one pseudophase is balanced; the mass and heat balance for both the gas and the solid phases is formulated in heterogeneous models. In this way, the temperature and the concentration differences between phases can be dissolved. Balances for the two-dimensional heterogeneous model of the reactor can be formulated in the following way: they must be formulated separately for the gas phase and separately for the solid phase.

And so for the gas phase:

$$\text{Mass balance}: \quad \frac{\partial(uc_i)}{\partial x} = D_{eff}\left(\frac{\partial^2 c_i}{\partial r^2} + \frac{1}{r}\frac{\partial c_i}{\partial r}\right) + k \tag{7}$$

$$\text{Energy balance}: \quad uc_t c_p \frac{\partial T}{\partial x} = \lambda\left(\frac{\partial^2 T}{\partial r^2} + \frac{1}{r}\frac{\partial T}{\partial r}\right) + k_h a\left(T_S^{pow} - T_C\right) \tag{8}$$

Reactor dimensions	Phase modeling	Kinetic approach	Software	Authors
1-D	Pseudohomogeneous	van Herwijnen	MATLAB	van Doesburg [28]
1-D	Pseudohomogeneous	Kopyscinski	Aspen Plus	Er-rbib et al. [29]
1-D	Pseudohomogeneous	Kopyscinski	ACM	Güttel [38]
2-D	Pseudohomogeneous	Xu and Froment	MATLAB	Schlereth and Hinrichsen [32]
3-D	Pseudohomogeneous	Measured (first order)	FEMLAB	Cao et al. [35]
0-D	Heterogeneous	Equilibrium	MATLAB	Rönsch et al. [36]
1-D	Heterogeneous	Xu and Froment	MATLAB	Matthischke and Rönsch [36]
1-D	Heterogeneous	Xu and Froment	MATLAB	Parlikkad et al. [37]
1-D	Heterogeneous	Xu and Froment	MATLAB	Schlereth and Hinrichsen [32]

Table 2.
Model characteristics and comparison of fixed-bed methanation reactors.

For the solid phase:

$$\text{Mass balance}: \quad k_m a\left(c_s^{surf} - c_i\right) + \rho_{bed} \sum_j^{N_r} \upsilon_{i,j} \eta_j r_j^{surf} = 0 \tag{9}$$

$$\text{Energy balance}: \quad k_h a\left(T_s^{surf} - T_i\right) + \rho_{bed} \sum_j^{N_r} \eta_j r_j^{surf} \left(-\Delta H_j\right) = 0 \tag{10}$$

Schlereth and Hinrichsen stated that the heterogeneous model should be employed, when the catalyst particles are large and have small pore radii or the methanation reactor operates with low Reynolds numbers. Miguel and Mendes [34] developed a kinetic model using industrial nickel-based catalyst. Model characteristics and comparison of fixed-bed methanation reactors was summarized in **Table 2**.

4. Summary and conclusions

In this section, mathematical and numerical models were presented and discussed. It is important to note that analysis of carbon monoxide and carbon dioxide methanation kinetics running in industrial conditions leads to the following conclusions:

- CO methanation rate is about twice as high as methanation CO_2 at the same concentration and temperature.
- The value of exponent at CO concentration, which is dependent on its value and on the range of small concentrations (below concentration around 500 ppm), tends to 1. For higher concentrations, the exponent value is about 0.7.
- Exponent at CO_2 concentration has a constant value of about 0.65 in the entire range of CO_2 concentration tested.
- Values of exponents tend to decrease at the lowest measuring temperature.
- CO methanation rate is not dependent on CO_2 concentration.
- The rate of CO_2 methanation decreases markedly with increasing concentration; thus, the catalyst activity in CO_2 methanation has decisive technological importance.

Author details

Artur Wodołażski
Central Mining Institute, Katowice, Poland

*Address all correspondence to: awodolazski@gig.eu

References

[1] Liu Z, Chu B, Zhai X, Jin Y, Cheng Y. Total methanation of syngas to synthetic natural gas over Ni catalyst in a micro-channel reactor. Fuel. 2012;**95**:599

[2] Kho ET, Jantarang S, Zheng Z, Scott J, Amal R. Harnessing the beneficial attributes of ceria and titania in a mixed-oxide support for nickel-catalysed photothermal CO_2 methanation. Engineering. 2017;**3**: 393-401

[3] Martin JW. Liquid-Phase Methanation/Shift Process Development, Report. Chem Systems Inc; 1982

[4] Beuls A, Swalus C, Jacquemin M, Heyen G, Karelovic A, Ruiz P. Methanation of CO_2: Further insight into the mechanism over Rh/γ-Al_2O_3 catalyst. Applied Catalysis B, Environmental. 2012;**113-114**:2

[5] CO_2-SNG Internal Report-TAURON. TAURON Production SA; 2015

[6] Więcław-Solny L, Tatarczuk A, Krótki A, Stec M. The technological research progress of amine-based CO_2 capture. Energy Politics. 2013;**16**(4): 229-240

[7] Müller K, Fleige M, Rachow F, Schmeißer D. Sabatier based CO_2-methanation of flue gas emitted by conventional power plants. Energy Procedia. 2013;**40**:240-248

[8] Ahern EP, Deane P, Persson T, Gallachóir BO, Murphy JD. A perspective on the potential role of renewable gas in a smart energy island system. Renewable Energy. 2015;**78**: 648-656

[9] Schaaf T, Grünig J, Schuster MR, Rothenfluh T, Orth A. Methanation of CO_2-storage of renewable energy in a gas distribution system. Energy, Sustainability and Society. 2014;**4**:1-14

[10] Rostrup-Nielsen JR, Pedersen K, Sehested J. High temperature methanation sintering and structure sensitivity. Applied Catalysis A: General. 2007;**330**:134

[11] Zhao A, Ying W, Zhang H, Ma H, Fang D. Ni/Al_2O_3 catalysts for syngas methanation: Effect of Mn promoter. Journal of Natural Gas Chemistry. 2012; **21**:170

[12] Wu H, Liu JX, Liu H, He D. CO_2 reforming of methane to syngas at high pressure over bi-component Ni-Co catalyst: The anti-carbon deposition and stability of catalyst. Fuel. 2019;**235**: 868-877

[13] Horlyck J, Lawrey C, Lovell EC, Amal R, Scott J. Elucidating the impact of Ni and Co loading on the selectivity of bimetallic NiCo catalysts for dry reforming of methane. Chemical Engineering Journal. 2018;**352**:572-580

[14] Karolyi J, Nemeth M, Evangelisti C, Safran G, Schay Z. Carbon dioxide reforming of methane over Nie In/SiO_2 catalyst without coke formation. Journal of Industrial and Engineering Chemistry. 2018;**58**:189-201

[15] Yamasaki M. Compositional dependence of the CO_2 methanation activity of Ni/ZrO_2 catalysts prepared from amorphous Ni-Zr alloy precursors. Applied Catalysis A: General. 1997;**163**:187

[16] Baumhakl C. Substitute natural gas production with direct conversion of higher hydrocarbons [PhD thesis]. Friedrich-Alexander-Universität Erlangen-Nürnberg; 2014

[17] Zhu L, Yin S, Yin Q, Wang H, Wang S. Biochar: A new promising catalyst

support using methanation as a probe reaction. Energy Science & Engineering. 2015;**3**:126-134

[18] Burger T, Koschany F, Thomys O, Kohler K, Hinrichsen O. CO_2 methanation over Fe- and Mn-promoted Co-precipitated Ni-Al catalysts: Synthesis, characterisation and catalysis study. Applied Catalysis, A: General. 2018;**558**:44e54

[19] Meerten RZC, Vollenbroek JG, de Croon MHJM, van Nisselrooy PFMT, Coenen JWE. The kinetics and mechanism of the methanation of carbon monoxide on a nickel-silica catalyst. Applied Catalysis. 1982;**3**:29-56

[20] Kopyscinski J, Schildhauer TJ, Biollaz SMA. Fluidized-bed methanation: Interaction between kinetics and mass transfer. Industrial and Engineering Chemistry Research. 2010;**50**:2781-2790

[21] Chein RY, Chen WY, Yu CT. Journal of Natural Gas Science and Engineering. 2016;**29**:243-251

[22] Hayes RE, Thomas WJ, Hayes KE. A study of the nickel-catalyzed methanation reaction. Journal of Catalysis. 1985;**92**:312-326

[23] Karolyi J, Nemeth M, Evangelisti C, Safran G, Schay Z, Horvath A, et al. Carbon dioxide reforming of methane over Ni/SiO_2 catalyst without coke formation. Numerical simulation of carbon dioxide methanation reaction for synthetic natural gas production in fixed-bed reactor. Journal of Industrial and Engineering Chemistry. 2018;**58**: 189-201

[24] Zhang J, Bai Y, Zhang Q, Wang X, Zhang T, Tan Y, et al. Low-temperature methanation of syngas in slurry phase over Zr-doped $Ni/c\text{-}Al_2O_3$ catalysts prepared using different methods. Fuel. 2014;**132**:211-218

[25] Sehested J, Dahl S, Jacobsen J, Rostrup-Nielsen JR. Methanation of CO over nickel: Mechanism and kinetics at high H_2/CO ratios. The Journal of Physical Chemistry B. 2005;**109**: 2432-2438

[26] Fujita S-I, Terunuma H, Nakamura M, Doi T, Takezawa N. Mechanisms of methanation of CO and CO_2 over Ni. Industrial and Engineering Chemistry Research. 1991;**30**:1146-1151

[27] Dissinger GR, Johnson SH, Stein FP. Dynamic simulation of methanation reactors, interstage heat exchanger, and controllers. Chemical Engineering Communications. 2007;**4**:577-591

[28] Doesburg H, De Jong WA. Transient behaviour of an adiabatic fixed-bed methanator—I: Experiments with binary feeds of CO or CO_2 in hydrogen. Chemical Engineering Science. 1976;**31**: 45-51

[29] Er-rbib H, Bouallou C. Modelling and simulation of methanation catalytic reactor for renewable electricity storage. Chemical Engineering Transactions. 2013;**35**:541-546

[30] Xu J, Froment GF. Methane steam reforming, methanation and water-gas shift: 1. Intrinsic kinetics. AICHE Journal. 1989;**35**:88-96

[31] Khorsand K, Marvast MA, Pooladian N, Kakavand M. Modeling and simulation of methanation catalytic reactor in ammonia unit. Petroleum and Coal. 2007;**49**:46-53

[32] Schlereth D, Hinrichsen O. A fixed-bed reactor modeling study on the methanation of CO_2. Chemical Engineering Research and Design. 2014; **92**:702-712

[33] Bader A, Bauersfeld S, Brunhuber C, Pardemann R, Meyer B. Modelling of a chemical reactor for simulation of a

methanisation plant. In: Proceedings, 8th International modelica conference; Dresden. 2011

[34] Miguel CV, Mendes A, Madeira LM. Intrinsic kinetics of CO_2 methanation over an industrial nickel-based catalyst. Journal of CO_2 Utilization. 2018;**25**: 128-136

[35] Cao C, Wang Y, Holladay JD, Jones EO, Palo DR. Design of micro-scale fuel processors assisted by numerical modeling. AICHE Journal. 2005;**51**: 982-988

[36] Rönsch S, Köchermann J, Schneider J, Matthischke S. Global reaction kinetics of CO and CO_2 methanation for dynamic process modelling—Review, selection, empirical adaption and application to a one-dimensional dynamic model with commercial Ni/ Al_2O_3 catalysts. Chemical Engineering and Technology. 2015

[37] Parlikkad NR, Chambrey S, Fongarland P, Fatah N, Khodakov A, Capela S, et al. Modelling of fixed bed methanation reactor for syngas production: Operating window and performance characteristics. Fuel. 2013; **107**:254-260

[38] Güttel R. Study of unsteady-state operation of methanation by modelling and simulation. Chemical Engineering and Technology. 2013;**36**:1675-1682

Chapter 6

Advances in Selective Oxidation of Methane

Saeed Alshihri and Hamid Almegren

Abstract

Selective oxidation of methane is one of the most challenging reactions in catalysis. Methane is a very stable molecule and requires high energy to be activated. Different approaches of single step methane conversion have been suggested to overcome this challenge. However, the current commercial process of methane conversion to methanol is via the indirect way, in which methane is first converted to synthesis gas in highly intensive energy step, and synthesis gas is then converted into methanol. The first step is responsible for 60% of the capital cost of the plant. There are enormous researches that have been conducted in a direct way and some good results have been achieved. This chapter will summarize the recent advances in the direct selective oxidation of methane to methanol.

Keywords: direct methane conversion, natural gas, oxygenates, catalysis, methanol

1. Introduction

The global consumption of energy and hydrocarbon-related commodities will continue to increase as the world population increases. The three sources of energy oil, gas, and coal are still dominating over 80% of the global energy matrix. However, natural gas is considered as the bridge fuel between the fossil fuel of today and the renewable fuel of tomorrow. It is cheap, more abundant than oil, and has lower CO_2 emissions compared with oil and coal. These factors place natural gas, and by extension methane, as a principal candidate for replacing petroleum as a chemical feedstock and addressing various environmental issues. Natural gas is a flammable substance obtained from oil or gas fields and coal mines. At present, the confirmed natural gas reserves have a total volume of 187 trillion metric, of which 24.8% is found in the Middle East, 30.4% in Europe and Eurasia, 8.4% in the Asia Pacific region, 7.5% in Africa, 6.8% in North America and 4.1% in Middle and South America [1–5]. Natural gas is typically used as a fuel for power generation and for domestic heating. In 1971, global primary energy consumption was based on 48% oil, 29% coal and 18% natural gas. However, in 2015, the consumption of 13.1 billion tonnes (oil equivalent) of fuel was based on 33% oil, 30% coal and 24% natural gas [1], reflecting a shift from oil to natural gas. This transition from oil to natural gas consumption is expected to gradually increase until 2035 [1–3].

Natural gas resources are located in remote areas, and its utilisation is affected by high transportation costs. Therefore, conversion of natural gas to high value chemicals is the most promising solution. Methane and ethane are the main components of natural gas; they are stable and have no functional group, magnetic moment or polar distribution to facilitate chemical attacks. The C-H bonds of these light hydrocarbons are strong and require high reaction conditions to be activated.

One of the most challenging processes in the chemical industry is the conversion of natural gas or methane to methanol, which is an important intermediate source of energy in our daily lives. Methanol can be used as a convenient energy storage material, a fuel, and a feedstock to synthesise hydrocarbons which mankind get from fossil fuel nowadays [2, 3] One of the importances of methanol comes from its direct use as a fuel or blending with gasoline to improve the octane number although it has half the volumetric energy density (15.6 MJ/L) relative to gasoline (34.2 MJ/L) and diesel (38.6 MJ/L) [4–6]. There had been 15,000 methanol-powered cars during the 1990s granted by the Environmental Protection Agency (EPA), but the use was discontinued due to an increased natural gas price [7]. Methanol is also a key feedstock for chemical manufacturing. The most major derivatives from methanol are formaldehyde, acetic acid, methyl tertiary butyl ether (MTBE) and dimethyl ether (DME). In recent years, methanol to hydrocarbons (MTH) research has been growing rapidly including methanol-to-gasoline (MTG) and methanol-to olefins (MTO) technology [8–10].

In industries, an indirect route for the conversion of natural gas to methanol is used. In this reaction, methane is first converted to synthesis gas by steam reforming, and the synthesis gas is then converted to methanol. However, the production of syngas is an energy-intensive process, which is operated between 800 and 1000°C, and more than 25% of the feed (natural gas) has to be burned to provide the heat of reaction. The direct conversion of methane to methanol in a single step without going through the reforming step is a desired alternative to the current technology [2, 4, 5]. In spite of the fact that there are no actual plants yet for the process of direct methane to methanol (DMTM), previous experimental and theoretical works have demonstrated the feasibility of this route [2, 4, 5]. Here, this chapter will mainly focus on the recent efforts on the direct conversion of methane to oxygenates.

2. Conventional catalytic approach to convert methane to oxygenates

2.1 Gas phase reaction based on homogeneous radical mechanism

This reaction is a free radical conducted under high temperature and pressure. The thermodynamic and kinetics studies identified the partial oxidation of methane as the rate limiting step due to the formation of methyl radical [7, 11]. Many studies with different oxidants have been conducted in this route. Babero et al. studied the partial oxidation of methane at 500 C temperature using nitrogen oxide as an oxidant [12]. Another study compared between oxygen and nitrogen oxide for the partial oxidation of methane in the gas phase [13]. The effect of adding small quantities of hydrocarbons such as ethane was investigated to promote the activation of methane and increase the selectivity of methanol [14]. Pressure is one of the most important factors which has a pronounced effect on the selectivity of methane oxidation. Dozens of studies have been performed in attempts to promote the selectivity toward oxygenates using high pressures and temperatures [8, 9, 15]. The results of these studies show that a conversion of 5–10% and a methanol yield of 30–40% can be achieved at a temperature of 723–773 K and pressures of 30–60 bar in the gas phase reaction. There are several works that investigated the reactor design and modifications. Zhang et al. designed a new tubular reactor based on quartz-line and stainless-steel line. The reaction was conducted at a temperature of 430–470°C and 5.0 Mpa pressure, and a high yield of methanol was obtained [9].

The methane conversion to methanol was also conducted in the absence of catalysts at high reaction conditions. Methanol yields as high as 7–8% are obtained in the absence of catalysts operating at 350–500°C and 50 bar [10, 11, 16]. As reactor inertness is essential for obtaining good selectivity to methanol, the feed gas should be isolated from the metal wall by using quartz and Pyrex glass-lined reactors [17]. A typical experimental conversion-selectivity plot for the gas-phase partial oxidation of methane is shown in **Figure 1** [18]. This plot ably demonstrates that any improvement in the direct conversion of methane to methanol must come from the enhancement of selectivity without reducing the conversion per pass. The Huels process uses cold-flame burners operating at 60 bar, with a selectivity of 71% to methanol and 14% to formaldehyde, and a recycle ratio of 200 to 1 [8].

The suggested mechanism for the direct conversion of methane to methanol via homogeneous gas phase reactions is shown in **Figure 2** [19].

2.2 A low temperature catalytic route involving homogeneous and heterogeneous catalysis

At moderate conditions, catalysts play an important role in the partial oxidation conversion of methane to methanol in terms of controlling the selectivity of the desired yield. Several catalysts have been investigated at 1 atm and mild temperatures.

In homogeneous systems, in the early 1970s, it was shown that methane could be converted to methanol by Pt(II) and Pt(IV) complexes because these complexes do not oxidise CH_3OH to CO_x [20]. Since that breakthrough, several oxidation catalysts based on Pt(II), Pd(II) and Hg(II) salts have been proven to functionalize C-H bonds [21, 22], leading to good yields of partially oxidised products (Eq. (1)). For example, [(2,2′-bipyrimidine)$PtCl_2$] catalyses the selective oxidation of CH_4 in fuming sulphuric acid to give methyl bisulphate in a 72% one-pass yield at 81% selectivity based on methane. Methyl bisulphate is then hydrolysed to methanol (Eq. (2)).

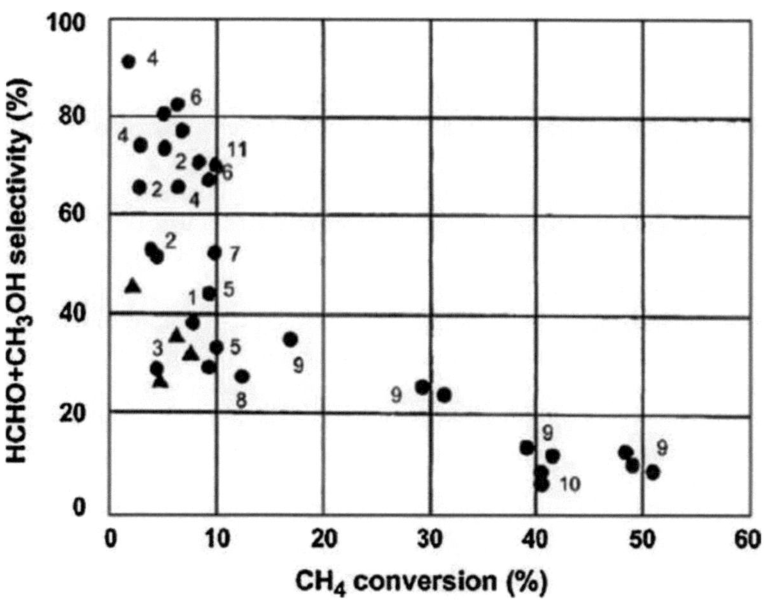

Figure 1.
Homogeneous gas-phase partial oxidation of methane from several studies: (1) Lott and Sliepcevich; (3) Tripathy; (4) Brockhaus; (6) Hunter; (8) Rytz and Baiker [18].

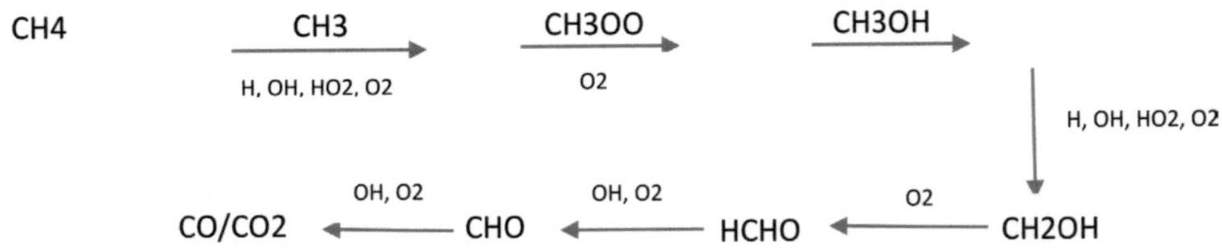

Figure 2.
Schematic diagram of the methane conversion via homogeneous gas phase reaction.

$$CH_4 + 2H_2SO_4 \rightarrow CH_3OSO_3 + SO_2 + 2H_2O \quad (1)$$

$$CH_3OSO_3 + H_2O \rightarrow CH_3OH + H_2SO_4 \quad (2)$$

The major drawbacks of the liquid phase include not only the difficulty of separating the methanol product from the solvent but also solvents such as sulphuric acid need expensive corrosion-resistant materials and periodic regeneration of the consumed H_2SO_4. A new class of solid catalyst based on immobilised complexes has recently been reported for the direct low-temperature oxidation of methane to methanol [23]. This solid catalyst has a covalent triazine-based framework (CTF) with numerous accessible bipyridyl structure units, which should allow the coordination of platinum (**Figure 3a** and **b**) [23]. The performance of these new catalysts showed that the activity is maintained throughout several cycles, and selectivity for methanol formation above 75% could be reached.

In nature, methane monooxygenase enzymes (MMO) transform CH_4 to CH_3OH in water under ambient conditions [11]. A number of metal complexes have been proposed to mimic the chemistry of these enzymes [11, 24, 25], but the systems which generate active oxygen species capable of converting CH_4 to CH_3OH are yet to be created. In contrast to organometallic CH_4 activation, MMO proceeds via a different mechanism by creating a very strong oxidising di-iron species able to attack a C-H bond in CH_4. An essential feature of MMO is an active site containing two iron centres [11]. Metallophthalocyanines (MPc), and more specifically iron phthalocyanines (FePc), are good catalysts for clean oxidation processes. More specifically, FePc supported in μ-oxo dimeric form (Fe-O-Fe fragment) has better catalytic properties in CH_4 conversion in the presence of hydrogen peroxide as an oxidant than its monomer counterpart (FePc). The heterolytic cleavage of the O-O bond in the $Fe^{IV}NFe^{III}OOH$ complex and the formation of very strong oxidising $Fe^{IV}NFe^{V} = O$ species are favoured in the presence of acid by the protonation of peroxide oxygen [11, 24, 25]. A new oxidation mechanism based on the use of metal clusters to harness the 'singlet oxene', the most reactive form of the oxygen atom, has recently been proposed [11, 26]. In this proposal, the key to oxygen insertion is a complex containing three copper atoms, in which the atomic charges vary. By synthesising a series of ligands to complex three copper atoms, mimicking the likely structure of the active site in pMMO, facile O-atom insertion into C-C and C-H bonds has been demonstrated in a number of simple organic substrates under ambient conditions of temperature and pressure. The ligands were designed to form the proper spatial and electronic geometry to harness a 'singlet oxene' [11, 25, 26]. It has been shown that the activity for methanol synthesis is 5 mol (CH_3OH) kg (catalyst)$^{-1}$ h^{-1} for sMMO as a complete enzyme with NADH present and this result represents the bench-market by which MMO catalysts should be judged. However, when NADH cofactor is removed, H_2O_2 can be used as the terminal oxidant with the enzyme but the catalytic activity decreases to 0.076 mol (CH_3OH) h^{-1} kg (MMOH)$^{-1}$ [27].

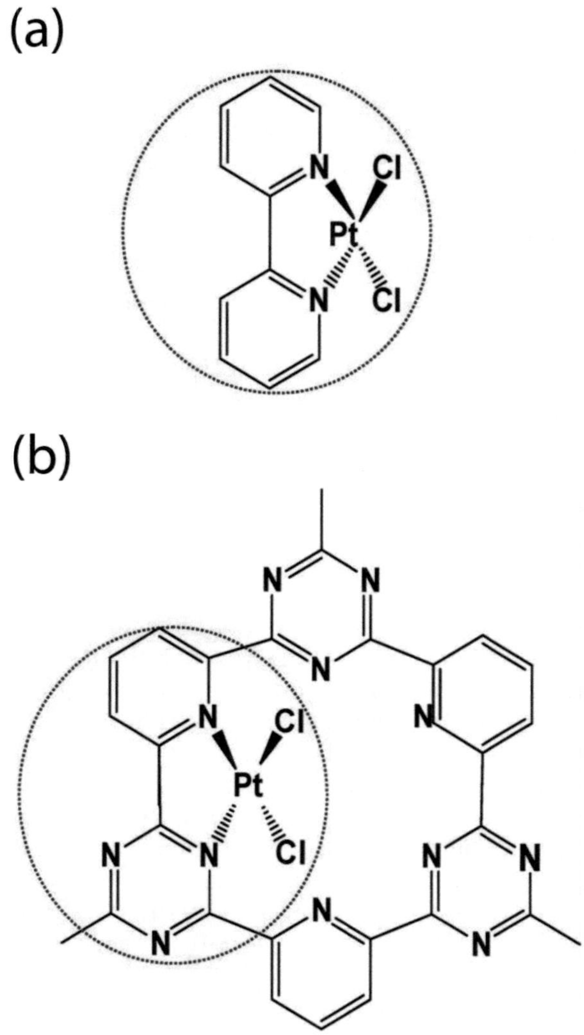

Figure 3.
(a) Bipyrimidyl Pt(II) complex used in the oxidation of methane to methyl bisulphate in concentrated sulphuric acid. (b) Covalent triazine-based framework (CTF) with numerous accessible bipyridyl structure units which are suitable to coordinate the Pt(II) complex.

In heterogeneous catalytic systems, many attempts have been conducted for the oxidation of methane. In most cases, SiO_2 was used as a support with different metals, and O_2 as oxidant. It was claimed that [28, 29] with a similar condition, HCHO might be produced with one-pass yield from 0 to 4%. However, high one pass HCHO yields were reported in some publications, but the results were not confirmed by other groups. It is stated that the results by different researchers have always been quite different, and some of them were even contradictory to one other [25, 30]. For instance, in one published work, a high selectivity (90%) to oxygenates (CH_3OH + HCHO) was obtained at CH_4 conversion of 20–25% at 873 K in an excess amount of water vapour over MoO_3/SiO_2 catalysts prepared by a sol-gel method [31]. Another work conducted the experiments under similar reactions with MoO_3/SiO_2 catalyst prepared by a similar method, but the yield of oxygenates was not greater than 4% [25]. Another example of contradiction showed that by using N_2O as an oxidant, CH_3OH could be achieved with a noticeable selectivity in the presence of H_2O over MoO_3/SiO_2 [32]. Results published by other groups used similar catalytic systems, but the results showed no detectable formation of CH_3OH [33, 34]. Metal-containing zeolites such as Co, Cu, and Fe have been studied at low temperature in batch mode [35, 36]. The direct conversion of methane to methanol over this metal-containing zeolite consists of three steps: (1) formation of active species by calcinations in air, (2) reaction of the active species with methane at low temperature and (3) extraction of methanol, using a polar protic solvent [37]. However, these catalytic systems are not yet a continuous process as an extraction procedure for methanol is required [36, 37]. A series of catalysts based on MoO_3 and WO_3 were studied, and the WO_3-based catalysts were less effective for the production of methanol. The Ga_2O_3/MoO_3 catalyst showed the maximum methanol yield [38].

In a series of seminal publications, Hutchings and co-workers demonstrated that Fe-ZSM-5- and Fe-Cu-ZSM-5-based zeolites and Au-Pd supported on titania could activate methane at temperatures under 100°C using aqueous hydrogen peroxide as the terminal oxidant [39, 40]. The initial product of reaction was shown to be methyl hydroperoxide which subsequently reacted to yield methanol and formic acid. **Figure 4** shows the time on line activity over Au-Pd/TiO_2.

The turnover frequencies based on Fe were high with albeit very low conversions. More recently, Al-Shihri et al. [41] showed that the reaction pathway of oxidation in aqueous hydrogen peroxide over ZSM-5 catalysts followed the reaction sequence $CH_4 \rightarrow CH_3OOH \rightarrow HCHO \rightarrow$ formic acid. Although at the reaction conditions used the formaldehyde was oxidised rapidly to formic acid, it was also converted to low molecular weight polyoxomethylene polymer. Similar findings were achieved in preliminary results using Au-Pd catalysts. However, the use of aqueous hydrogen peroxide to oxidise methane is unlikely to prove economic unless its parallel catalysed decomposition into oxygen and H_2O can be supressed. Thus, the development of a viable liquid phase process based on the use of aqueous hydrogen peroxide as the terminal oxidant would be challenging.

An attractive alternative approach is to couple in situ direct generation of hydrogen peroxide from hydrogen and oxygen with methane oxidation in a tandem process. Au-Pd catalysts have proved to be highly active for the direct hydrogen peroxide synthesis reaction and capable of enhancing the tandem catalytic oxidation of alcohols, especially using nanostructured oxide supports [42, 43]. However, while the production rate of hydrogen peroxide is high, the achievable

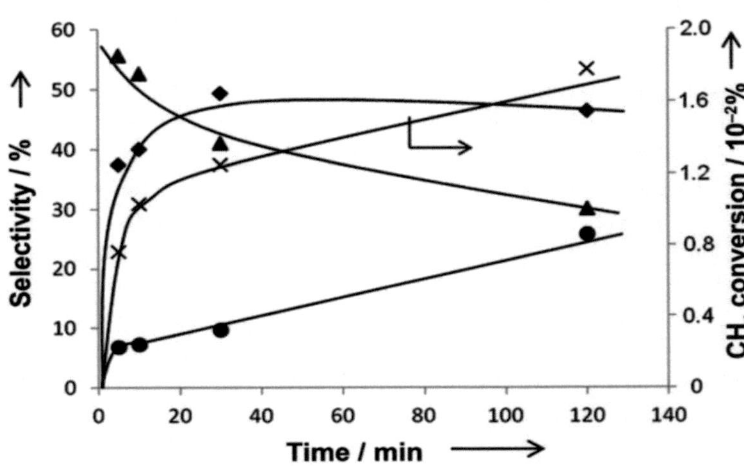

Figure 4.
Time on line activity reaction temperature: 50°C, [H_2O_2]: 5000 μmol, solvent: H_2O, 10 mL. Catalyst: 1.0×10^{-5} mol of metal, 28 mg 2.5 wt% Au-2.5 wt% Pd/TiO.

concentration in the liquid phase remains low due to the catalysed decomposition of the formed hydrogen peroxide. This means that a tandem process in the liquid phase is more likely to find application in the selective oxidation of high value chemicals. For direct selective oxidation of light hydrocarbons to oxygenated compounds, a gas phase continuous process based on the use of heterogeneous catalysts would be more attractive. In a tandem oxidation process, the oxidant would be oxygen, air or N_2O mixed with hydrogen to generate surface hydroperoxy in situ by the surface reaction of hydrogen-oxygen.

In a preliminary study, Al-Shihri et al. demonstrated that Au-Pd catalysts were able to catalyse the gas phase direct selective oxidation of methane at moderate conditions using the tandem synthesis of hydrogen peroxide from hydrogen-oxygen mixtures. The products of reaction were trapped and found to be methylhydroperoxide, polyoxomethylene, and a small amount of formic acid. Based on analogy with the liquid phase reaction sequence described above [41], the production of polyoxomethylene would be expected to involve initially the formation of formaldehyde as a reaction intermediate, although none was detected. This aspect is currently being investigated prior to publication of these exciting new results. The production of methyl hydroperoxide, formaldehyde and polyoxomethylene from methane is highly desirable. Polyoxomethylenes are valuable polymeric materials and also potential hydrogen storage materials; methyl hydroperoxide can be utilised to form methanol or react to form other compounds. In our preliminary study, the most promising Au-Pd catalysts were based on the use of nanostructured oxide supports. However, the catalysts were prepared by simple impregnation and were far from optimum in terms of metal particle dispersion and degree of Au-Pd alloy formation. These factors are important in the activity, selectivity, and maximising the selectivity in the use of the hydrogen, that is, avoiding direct combustion to water.

Shan et al. showed that mononuclear rhodium species supported by zeolite or titanium dioxide in aqueous solution can convert methane to methanol and acetic acid with high selectivity, using oxygen and carbon monoxide under mild conditions [44]. In a recent study, the direct conversion of methane to methanol was investigated using experimental and computational study. The results of this study showed that low Ni loadings on a $CeO_2(111)$ support can perform a direct catalytic cycle for the generation of methanol at low temperature using oxygen and water as reactants, with a higher selectivity than ever reported for ceria-based catalysts [45].

Gold-based catalysts have also shown interesting performance for the activation of C-H bond in alkane selective oxidation with dioxygen. A particular focus has been put on the synthesis of cyclohexanone and cyclohexanol. Zhao and co-workers [46] first applied gold catalysis in the activation of cyclohexane: Au/ZSM-5 and Au/MCM-41 favoured selectivity around 90% and conversions of 10–15% at 150°C, even though a loss in both activity and selectivity after their reuse is a drawback for industrial application. Two recent studies on the selective cyclohexane oxidation were performed by tailoring a supported gold on different materials, namely amorphous silica doped with titania and alumina prepared by a modified direct anionic exchange method [47].

In the direct gas phase oxidation of methane to methanol, no noble metal except Pd was investigated, and there was no promising results obtained when Pd was used and that might be due to the excessive interaction between Pd and the supports [25, 48]. Therefore, in order to overcome the extent of interaction between Pd and the support, and to increase selectivity toward methanol, bimetallic systems seem to be a more promising solution. Great success has been achieved in a variety of catalytic processes by combining two metallic elements in bimetallic catalysts, such as the platinum-iridium (Pt-Ir) system for petroleum reforming, platinum-tin (Pt-Sn) for alkane dehydrogenation, the nickel-gold (Ni-Au) system for steam reforming of alkanes, and the palladium-gold (Pd-Au) for selective oxidations [49].

2.3 Challenges in technologies for the conventional methods

The industrial production of methanol is executed via indirect way, in which methane is first converted to synthesise gas in highly intensive energy step. The synthesis gas is then converted to methanol. The intensive energy synthesis gas step occurs in operational plant at pressure range between 200 and 600 psi and temperature range between 700 and 1000°C [50]. This step is responsible for 60% of the capital cost of the plant. Therefore, the direct conversion of methane to methanol is highly desirable. Several approaches have been investigated and reported; however, no breakthrough has been achieved yet.

The homogeneous gas phase partial oxidation has the potential to replace the industrial method. In a technical evaluation study of this method, it was shown that a methanol selectivity of over 70% at 15% methane conversion can be achieved. However, the low conversion of methane per pass and relatively low methanol selectivity is still observed in most of the academic reports [51, 52]. The problem is due to kinetic and thermodynamic reasons [53]. This way requires a pressure of around 10 atm and temperature (1000°C) to activate methane and convert it to methanol with reasonable selectivity. The C-H bond in methane (440 kJ/mol) is stronger than the same bond in methanol (389 kJ/mol). That means at high temperatures, the methanol is more reactive than methane, which might lead to the decomposition of methanol to low grade product [52, 54–56]. In addition, the gas phase homogeneous is a free radical reaction, which means that it is difficult to control the process on the large scale [51, 54, 57].

The catalysts play an important role in activating methane at low reaction conditions and produce methanol with low byproducts [58, 59]. Two advantages of this method are the reduction of energy consumption used for methane conversion to methanol, and the low concentration of CO_2 produced in this process [58, 59]. The important factor in this process is to find catalysts that could activate methane at moderate conditions and convert it selectively to methanol. Although intensive work has been reported, no catalytic system has achieved the target conversion and selectivity.

A low temperature homogeneous catalyst in solutions is another way to convert methane to methanol at low temperatures. However, two challenges of this method is first the introduction of the catalysts with reasonable reactivity and selectivity that also tolerates oxidising and protic conditions [11]. The second challenge is the use of acid as a solvent such as sulphuric acid, which is applied in many studies. The main disadvantage of using sulphuric acid as solvent is the difficulty in separating the methanol from the solvent. Moreover, the acid might corrode and poison the catalysts through the reaction [11].

In nature, methane monooxygenases (MMOs) demonstrate high activity for methanol synthesis with a production rate of 5 mol (CH_3OH) kg (catalyst)-1 h^{-1} at ambient conditions. However, this method is still not practical yet due to the difficulty in purifying these proteins and the further oxidation reaction of methanol to formaldehyde.

3. Unconventional approach to convert methane to oxygenates

3.1 Conversion of methane to methanol via plasma technologies

Plasma can be used in many applications including oxidation of methane to methanol [60]. In plasma, the oxidation of methane to methanol can be conducted under atmospheric gas pressure. Plasma is often referred to as the fourth state of matter, and it includes several components: positive ions, negative ions, electrons and neutral species. Plasma technology can be classified into thermal plasma and non-thermal plasma [61]. Thermal plasma can be described as a gas consisting of electrons, highly excited atoms, ions, radicals, photons and neutral particles, while electrons that have much higher energy than other surrounding particles populate non-thermal plasma. Okazaki et al. [62] reported that the conversion of methane to methanol was achieved using non-equilibrium plasma chemical reactions under atmospheric pressure by ultra-short pulsed barrier discharge in an extremely thin glass tube reactor. Various designs for plasma reactors for the oxidation of methane have been proposed to enhance the selectivity toward methanol. For example, in thermal plasma reactors, the dielectric barrier discharge (DBD) reactor was used for the synthesis of methanol from methane. This reactor was able to reduce the required temperature and pressure needed [63]. Another reactor is a new non-thermal discharge micro-reactor, which is used for a single-step, non-catalytic, direct and selective synthesis of methanol via methane partial oxidation at room temperature [64]. The non-thermal plasma can be developed by integrating the reactor with catalyst to improve the activity and selectivity of methane oxidation. In a recent study, the Cu-doped Ni was loaded on CeO_2, which led to enhance the selectivity of methanol until 36% [65]. In another study, multicomponent catalysts were combined with plasma in two different approaches, in-plasma catalysis (IPC) and post-plasma catalysis (PPC), for achieving high levels in both methane conversion and aimed methanol selectivity through the synergetic effect of the Fe_2O_3¯CuO/γ-Al_2O_3 catalyst [66].

3.2 Methane oxidation to methanol using photocatalysts

The photocatalytic process is a photochemical reaction that is carried out with external energy provided by ultraviolet light radiation that has energy equal to or greater than the energy band gap of a semiconductor. Several of oxidation and reduction processes are involved in the photo-generated electron and hole. TiO_2 catalysts have been used as semiconductor photocatalysts for a wide range of environmental applications [67]. In addition, tungsten oxide (WO_3) is also a good photocatalyst due to its high chemical stability in aqueous solution under acidic conditions in the presence of an oxidising agent [68]. For example, one study demonstrated that the WO_3 photocatalyst produced hydroxyl radicals that react with a methane molecule to produce a methyl radical, which promote the formation of methanol [69]. Another study [70] investigated different experimental parameters for the methane conversion such as catalyst concentration, laser power, laser exposure time, effects of free radical generator (H_2O_2) and electron capture agent (Fe^{3+}), using visible laser light. Also, this study examined the comparison between WO_3 and TiO_2, and it was found that the WO_3 showed the highest methane conversion [70, 71]. A recent work has studied the introduction of some electron scavengers such as (Fe^{3+}, Cu^{2+}, and Ag_+) and H_2O_2 species to the WO_3 catalyst to enhance the selectivity of methanol. They found that WO_3/Fe^{3+} is the most active catalyst with a methanol selectivity of 58.5% [68]. Another photocatalyst for the methane oxidation to methanol is vanadium oxide supported by MCM-41. Nitric oxide (NO) was used as an oxidant for the oxidation of methane under UV irradiation at 295 [72]. **Figure 5** shows an example of methane conversion to methanol via photocatalysis.

3.3 Conversion of methane to methanol using supercritical water

The supercritical water oxidation (SCWO) is a reaction that occurs in water at elevated temperatures and pressures above the thermodynamic critical point of the

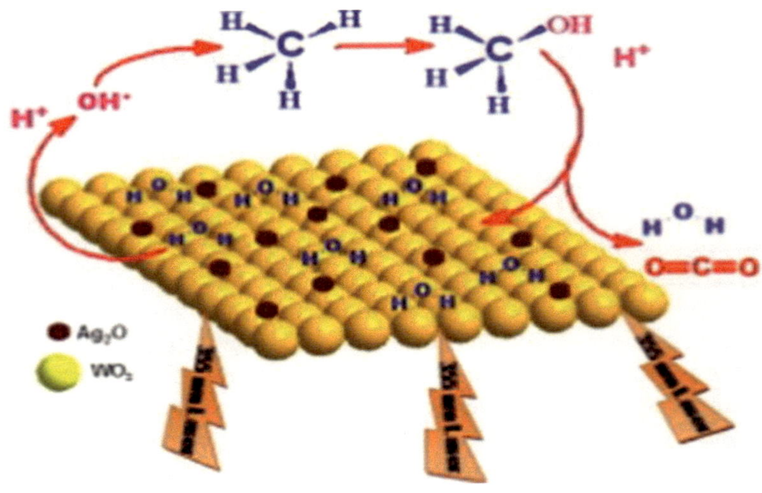

Figure 5.
Graphical representation of reaction of conversion of methane to methanol via photocatalysts [73].

mixture. Under the supercritical fluid conditions, the properties of water such as viscosity and dielectric constant can be adjusted between high gas-like diffusion rates and high liquid-like collision rates by varying pressure and temperature [60]. The catalytic oxidation of methane was examined over Cr_2O_3 under supercritical water conditions, and it was found that this catalytic system under supercritical conditions enhances the conversion rate of ethane and promotes the selectivity of methanol [74]. Another study investigated the isothermal conditions with a laminar reactor in SCWO for the direct partial oxidation of methane to methanol. They achieved a methanol selectivity of 35% at a conversion of 3% at temperatures of 400–410°C [75]. Savage PE et al., [76] have examined two types of reactors, glass-lined reactors and stainless steel reactors. A parametric study has been conducted using both reactors, and the glass lined reactor showed higher conversion of methane to methanol.

3.4 Conversion of methane to methanol using membrane technology

Membrane technology has been used for methane conversion to methanol using membrane reactor at moderate conditions. The advantage of using a membrane reactor is the fact that it can perform two functions at once, reaction and separation. The membrane can be classified based on the type of materials and porosity. The membrane can be made either by polymeric or organic materials with different porosity [60]. The organic membrane has advantages over the polymer in terms of the tolerance to chemical and temperature effects. Moreover, the organic membrane is mainly composed of metallic or ceramic materials and has greater physiochemical stability. Two research works studied the methane oxidation to produce methanol using Methylosinus trichosporium OB3b with a high concentration of Cu^{2+} and they found that the optimization of the conversion rate was positively affected by several parameters including the temperature, pH and concentrations of sodium formate, phosphate buffer and cyclopropanol [77, 78]. In another study, the methane oxidation was carried out using a membrane reactor where the methane and oxygen were introduced by two separate dense silicone tubes. A high methanol production of 1.12 g/L and 60% methane conversion were reported [79].

3.5 Challenges of unconventional ways to convert methane to methanol

The unconventional technologies such as plasma, photocatalysts, supercritical and biological are long-term processes, and still away from practical. The methane oxidation under plasma conditions is considered as a clean method as there are no harmful emissions produced such as CO_2 and CO [80]. The plasma reactor is simple, benign and cheap. However, the productivity of methanol is low due to the limitation of methane solubility in the reaction medium at ambient conditions.

Photocatalysis technology is an attractive way to convert methane to methanol, as the basic requirement for this method is the use of three abundant reactants of light, water and methane. Despite the two decades of work on photocatalysis, the selectivity of methanol is still low [68, 70].

Supercritical water oxidation is an efficient process to treat a variety of hazardous and non-hazardous wastes. However, there are some factors that limit the application of this technology in methane oxidation such as the complication of the reactor design, the high temperature used and the high corrosion rate when using halogens such as chlorine for some waste treatment [60, 81].

The use of membrane technology for methane conversion to methanol is feasible due to the advantage of the effective separation of methane and methanol. However, some challenges still exist for large scale application: first, the relatively high energy requirement for large scale plant, second, the low tolerance of polymer-based membrane to high temperatures and chemicals, third, the high conversion of methane will produce different organic compounds, and that might cause swelling or breakage of the membrane [60, 81].

4. Summary

In this chapter, we discussed the utilisation of methane as the main component of natural gas that can be converted to methanol. We discussed various processes that can activate methane and convert it in a single step to methanol, including their feasibility, recent progress and challenges associated with the conventional and unconventional methods. We showed that these processes have advantages and disadvantages. However, most of them suffer from the low yield of methanol. The unconventional methods are long-term processes and still far away from practical. The low temperature route using heterogeneous catalysts has a great potential and can be alternative to the current industrial process as some catalytic systems were shown to activate methane at moderate reaction conditions using different oxidants. Nevertheless, the selectivity toward methanol is still low. Therefore, more effort is needed to design and synthesise robust and cheap catalysts that could convert methane directly and selectively to methanol using air as an oxidant in a continuous flow reaction system.

Author details

Saeed Alshihri* and Hamid Almegren
National Petrochemical Technology Center, Materials Science Research Institute, King Abdulaziz City for Science and Technology (KACST), Riyadh, Saudi Arabia

*Address all correspondence to: alshihri@kacst.edu.sa

References

[1] British Petroleum. BP statistical review of world energy. 2016. https://www.bp.com/content/dam/bp-country/de_ch/PDF/bp-statistical-review-of-world-energy-2016-full-report.pdf

[2] Tomohiro Y, Yasushi S. Methane conversion using carbon dioxide as an oxidizing agent: A review. Fuel Processing Technology. 2018;**181**:187-198

[3] Edenhofer O et al. editors. IPCC Climate Change 2014: Mitigation of Climate Change. 2014

[4] Han B et al. A review of the direct oxidation of methane to methanol. Chinese Journal of Catalysis. 2016;**37**:1206-1215

[5] Arutyunov VS. Direct Methane to Methanol Foundations and Prospects of the Process. Amsterdam: Elsevier; 2014

[6] Wen-Chi L et al. The methanol economy: Methane and carbon dioxide conversion. Topics in Catalysis. 2018;**61**:530-541

[7] Yabe T, Sekine Y. Methane conversion using carbon dioxide as an oxidizing agent: A review. Fuel Processing Technology. 2018;**181**:187-198

[8] US Patent 7,097,675; 2006

[9] Zhang Q et al. Comparatively high yield methanol production from gas phase partial oxidation of methane. Applied Catalysis A: General. 2002;**224**:201-207

[10] Gesser HD, Hunter NR. A review of C-1 conversion chemistry. Catalysis Today. 1998;**42**:183-189

[11] Galvan A et al. Direct methane conversion routes to chemicals and fuel. Catalysis Today. 2011;**171**:15-23

[12] Baber JA et al. Breakthrough in C1-oxygenates production via direct methane oxidation. Chemical Communications. 2002;**11**:1184-1185

[13] Tabata K et al. Activation of methane by oxygen and nitrogen oxide. Catalysis Reviews: Science and Engineering. 2002;**44**:1-58

[14] Fujimoto K. New uses of methane. In: Studies in Surface Science and Catalysis, H.E.; 1994;**81**:73-84

[15] Holmen A. Direct conversion of methane to fuels and chemicals. Catalysis Today. 2009;**142**:2-8

[16] Gesser HD, Hunter NR, Prakash CB. The direct conversion of methane to methanol by controlled oxidation. Chemical Reviews. 1985;**85**:235-244

[17] Zhang Q, He D, Zhu Q. Direct partial oxidation of methane to methanol: Reaction zones and role of catalyst location. Journal of Natural Gas Chemistry. 2008;**17**:24-28

[18] Tabata K, Teng Y, Takemoto T, Suzuki E, Banares MA, Pena MA, et al. Activation of methane by oxygen and nitrogen oxides. Catalysis Reviews: Science and Engineering. 2002;**44**:1-58

[19] Brown MJ, Parkyns ND. Progress in the partial oxidation of methane to methanol and formaldehyde. Catalysis Today. 1991;**8**:305-335

[20] Goldshleger NF, Eskova VV, Shilov AE, Shteinman AA. Activation of saturated hydrocarbons by transition metal complexes, russian. Journal of Physical Chemistry A. 1972;**46**:785-786

[21] Periana RA et al. A mercury-catalyzed, High-yield system for the oxidation of methane to methanol. Science. 1993;**259**:340-343

[22] Periana RA et al. Platinum catalysts for the high-yield oxidation of methane to a methanol derivative. Science. 1998;**280**:560-564

[23] Palkovits R et al. Solid catalysts for the selective low-temperature oxidation of methane to methanol. Angewandte Chemie International Edition. 2009;**48**(37):6909-6912

[24] Anders H. Direct conversion of methane to fuels and chemicals. Catalysis Today. 2009;**142**(1-2):2-8

[25] Otsuka K, Wang Y. Direct conversion of methane into oxygenates. Applied Catalysis A: General. 2001;**222**(1-2):145-161

[26] Chen PP-Y et al. Facile O-atom insertion into C C and C H bonds by a trinuclear copper complex designed to harness a singlet oxene. Proceedings of the National Academy of Sciences. 2007;**104**(37):14570-14575

[27] Forde MM et al. Methane oxidation using silica-supported N-bridged di-iron phthalocyanine catalyst. Journal of Catalysis. 2012;**290**:177-185

[28] Kastanas GN, Tsigdinos GA, Schwank J. Selective oxidation of methane over vycor glass, quartz glass and various silica, magnesia and alumina surfaces. Applied Catalysis. 1988;**44**:33-51

[29] Alptekin GO et al. Methane partial oxidation by unsupported and silica supported iron phosphate catalysts: Influence of reaction conditions and co-feeding of water on activity and selectivity. Journal of Catalysis. 1999;**181**(1):104-112

[30] Zhang Q, He D, Zhu Q. Direct partial oxidation of methane to methanol: Reaction zones and role of catalyst location. Journal of Natural Gas Chemistry. 2008;**17**(1):24-28

[31] Sugino T et al. Partial oxidation of methane on silica-supported silicomolybdic acid catalysts in an excess amount of water vapor. Journal of Catalysis. 2000;**190**(1):118-127

[32] Liu HF, Liu RS, Liew KY, Johnson RE, Lunsford JH. Partial oxidation of methane by nitrous oxide over molybdenum on silica. Journal of the American Chemical Society. 1984;**106**:4117

[33] Barbaux Y, Elamrani A, Bonnelle JP. Catalytic oxidation of methane on MoO_3-SiO_2: Mechanism of oxidation with O_2 and N_2O studied by surface potential measurements. Catalysis Today. 1987;**1**(1-2):147-156

[34] Banares MA, Fierro JLG, Moffat JB. The partial oxidation of methane on MoO_3/SiO_2 catalysts: Influence of the molybdenum content and type of oxidant. Journal of Catalysis. 1993;**142**(2):406-417

[35] Beznis N, Weckhuysen B, Bitter J. Partial oxidation of methane over Co-ZSM-5: Tuning the oxygenate selectivity by altering the preparation route. Catalysis Letters. 2010;**136**(1):52-56

[36] Beznis NV et al. Oxidation of methane to methanol and formaldehyde over Co-ZSM-5 molecular sieves: Tuning the reactivity and selectivity by alkaline and acid treatments of the zeolite ZSM-5 agglomerates. Microporous and Mesoporous Materials. 2011;**138**(1-3, 183):176

[37] Cao G, Wang Y. Nanostructures and Nanomaterials. Synthesis, Properties and Applications. Singapore: World Scientific; 2011

[38] Taylor SH et al. The partial oxidation of methane to methanol: An approach to catalyst design. Catalysis Today. 1998;**42**(3):217-224

[39] Rahim MHA, Forde MM, Jenkins RL, Hammond C, He Q, Dimitratos N, et al. Oxidation of methane to methanol with hydrogen peroxide using supported gold–palladium alloy nanoparticles. Angewandte Chemie, International Edition. 2013;**52**:1318-1322

[40] Hammond C, Forde MM, Rahim MHA, Thetford A, He Q, Jenkins L, et al. Direct catalytic conversion of methane to methanol in an aqueous medium by using copper-promoted Fe-ZSM-5. Angewandte Chemie, International Edition. 2012;**51**:5129-5133

[41] Al-Shihri S et al. Selective oxidation of methane to methanol over ZSM-5 catalysts in aqueous hydrogen peroxide: Role of formaldehyde. ChemCatChem. 2017;**9**:1276-1283

[42] Torrente-Murciano L, He Q, Hutchings GJ, Kiely CJ, Chadwick D. Enhanced Au-Pd activity in the direct synthesis of hydrogen peroxide using nanostructured tubular titanate nanotube supports. ChemCatChem. 2014;**6**:2531-2534

[43] Torrente-Murciano L, Villager T, Chadwick D. Selective oxidation of salicylic alcohol to aldehyde with O_2/H_2 using Au-Pd on titanate nanotubes catalysts. ChemCatChem. 2015;7:925-927

[44] Shan J, Li M, Allard LF, Lee S, Flytzani-Stephanopoulos M. Mild oxidation of methane to methanol or acetic acid on supported isolated rhodium catalysts. Nature. 2017;**551**

[45] Lustemberg PG et al. Direct conversion of methane to methanol on Ni-ceria surfaces: Metal-support interactions and water-enabled catalytic conversion by site blocking. Journal of the American Chemical Society. 2018;**140**(24):7681-7687

[46] Torrente-Murciano L et al. Highly selective Pd/titanate nanotube catalysts for the double-bond migration reaction. Journal of Catalysis. 2007;**245**(2):272-278

[47] Della Pina C et al. Selective oxidation using gold. Chemical Society Reviews. 2008;**37**(9):2077-2095

[48] Juszczyk W et al. Pd-Au/SiO_2: Characterization and catalytic activity. Journal of Catalysis. 1995;**151**(1):67-76

[49] Shiju NR, Guliants VV. Recent developments in catalysis using nanostructured materials. Applied Catalysis A: General. 2009;**356**(1):1-17

[50] Arutyunov V. Low-scale direct methane to methanol-modern status and future prospects. Catalysis Today. 2013;**215**:243-250

[51] Zhang Q et al. Comparatively high yield methanol production from gas phase partial oxidation of methane. Applied Catalysis A: General. 2002;**224**:201-207

[52] Dallos CG, Kafarov V, Filho RM. Two dimensional steady-state model of the gas-solid-solid reactor: Example of the partial oxidation of methane to methanol. Chemical Engineering Journal. 2007;**134**:209-217

[53] Arutyunov VS. Recent results on fast flow gas-phase partial oxidation of loweralkanes. Journal of Natural Gas Chemistry. 2004;**13**:10

[54] Holmen A. Direct conversion of methane to fuels and chemicals. Catalysis Today. 2009;**142**:2-8

[55] Khirsariya P, Mewada RK. Single step oxidation of methane to methanol—Toward better understanding. Procedia Engineering. 2013;**50**:409-415

[56] Silva MJD. Synthesis of methanol from methane: Challenges and advances on the multi-step (syngas) and one-step

routes (DMTM). Fuel Processing Technology. 2016;**145**:42-61

[57] Mahammad SK et al. Catalytic non-thermal plasma assisted co-processing of methane and nitrous oxide for methanol production. Catalysis Today. 2015;**256**:102-107

[58] Zhang X, He D, Zhang Q, Xu B, Zhu Q. Catalytic conversion of methane to methanol over lanthanum cobalt oxide supported molybdenum based catalysts. Preprints of Papers-American Chemical Society. 2003;**48**(2):837

[59] Raja R, Ratnasamy P. Letter direct conversion of methane to methanol. Applied Catalysis A: General. 1997; 158:L7-15

[60] Reddy PVL, Kim KH, Song H. Emerging green chemical technologies for the conversion of methane to valu eadded products. Renewable and Sustainable Energy Reviews. 2013;**24**:578-585

[61] Roth JR. Industrial Plasma Engineering, Principles. Vol. 1. Bristol, UK: Institute of Physics Press; 1995

[62] Okazaki K et al. Direct conversion from methane to methanol for high efficiency energy system with exergy regeneration. Energy Conversion and Management. 2002;**43**:1459-1468

[63] Larkin D, Lobban L, Mallinson G. The direct partial oxidation of methane to organic oxygenates using a dielectric barrier discharge reactor as a catalytic reactoranalog. Catalysis Today. 2001;**71**:199-210

[64] Nozaki T et al. A single step methane conversion in to synthetic fuels using micro-plasma reactor. Chemical Engineering Journal. 2011;**166**:288-293

[65] Mahammad SK et al. Catalytic non-thermal plasma assisted coprocessing of methane and nitrous oxide for methanol production. Catalysis Today. 2015;**256**:102-107

[66] Chen L et al. Application of in-plasma catalysis and post-plasma catalysis form ethane partial oxidation to methanol over a Fe_2O_3-CuO/γ-Al_2O_3 catalyst. Journal of Natural Gas Chemistry. 2010;**19**:628-637

[67] Tahir M, Amin NS. Recycling of carbon dioxide to renewable fuels by photo-catalysis: prospects and challenges. Renewable and Sustainable Energy Reviews. 2013;**25**:560-579

[68] Villa K et al. Mesoporous WO_3 photocatalyst for the partial oxidation of methane to methanol using electrons cavengers. Applied Catalysis B: Environmental. 2015;**163**:150-155

[69] Taylor CE, Noceti RP. New developments in the photocatalytic conversion of methane tomethanol. Catalysis Today. 2000;**55**:259-267

[70] Gondala MA, Hameed A, Suwaiyan A. Photo-catalytic conversion of methane into methanol using visible laser. Applied Catalysis A: General. 2003;**243**:165-174

[71] Gondal MA et al. Photocatalytic transformation of methane into methanol under UV laser irradiation over WO_3, TiO_2 and NiO catalysts. Chemical Physics Letters. 2004;**392**:372-377

[72] Hu Y et al. Characteristics of the photo-catalytic oxidation of methane into methanol on V-containing MCM-41. Catalysis Letters. 2008;**124**:80-84

[73] Hameed A, Ismail IMI, Aslam M, Gondal MA. Photocatalytic conversion of methane into methanol: Performance of silver impregnated WO_3. Applied Catalysis A: General. 2014;**470**:327-335

[74] Dixon CN, Abraham MA. Conversion of methane to methanol by

catalytic supercritical water oxidation. Journal of Supercritical Fluids. 1992;**5**(4):269-273

[75] Lee JH, Foster NR. Direct partial oxidation of methane to methanol in super-critical water. Journal of Supercritical Fluids. 1996;**9**(2):99-105

[76] Savage PE et al. Methane to methanol in supercritical water. Journal of Supercritical Fluids. 1994;**7**(2):135-144

[77] Lee SG et al. Optimisation of methanol biosynthesis from methane using methylosinus trichosporium OB3b. Biotechnology letters. 2004;**26**:947-950

[78] Takeguchi M et al. Optimisation of methanol bio-synthesis by Methylosinus trichosporium OB3b: An approach to improve methanol accumulation. Applied biochemistry and biotechnology. 1997;**68**:143-152

[79] Duan C et al. High-rate conversion of methane to methanol by Methylosinus trichosporium OB3b. Bioresource technology. 2011;**102**:7349-7353

[80] Lemmens B et al. Assessment of plasma gasification of high caloric waste streams. Waste Management. 2007;**27**:1562-1569

[81] Introduction to membrane. http://www.separationprocesses.com/Membrane/MT_Chp01c.htm

Chapter 7

Partial Oxidation of Methane to Methanol on Cobalt Oxide-Modified Hierarchical ZSM-5

Yuni Krisyuningsih Krisnandi, Dita Arifa Nurani, Muhammad Reza, Bayu Adi Samodro, Suwardiyanto, Nirwan Susianto, Arbhyando T. Putrananda, Indah R. Saragi, Aminah Umar, Sung-Min Choi and Russell F. Howe

Abstract

Conversion of methane to more reactive compounds such as methanol has drawn attention for many years. Hierarchical ZSM-5 zeolite has been used as support of metal oxide catalyst to facilitate the partial oxidation of methane to methanol. The NaZSM-5 zeolite was synthesized hydrothermally using double-template techniques, in which tetrapropylammonium hydroxide (TPAOH) and polydiallyldiammonium chloride (PDDA) were used as primary and secondary templates, respectively. HZSM-5 was prepared through multiple $NH4^+$ exchange of NaZSM-5 followed by calcination. Co oxide-modified ZSM-5 (Co/NaZSM-5 and Co/HZSM-5) were prepared through impregnation method. Then, the zeolites were extensively characterized using scanning electron microscope (SEM), X-ray diffraction (XRD), AAS, Fourier transform infrared (FTIR), ^{27}Al solid-state NMR, microbalance, and surface area analysis. The catalytic test was performed in batch reactor, and the product was analyzed with GC-FID. Reaction condition and acidity of ZSM-5 as support catalyst were studied. As a result, when using Co/HZSM-5 as catalyst, percentage (%) yield of methanol was increased with longer reaction time. On the other hand, the percentage (%) yield decreased when Co/NaZSM-5 was employed. Introduction of trace amount of oxygen to the gas mixture showed different results. Furthermore, the prospect of synthesis of ZSM-5 using natural resources and using biogas are also explored.

Keywords: hierarchical zeolite, ZSM-5, partial oxidation, methane, methanol

1. Introduction

Methane is a greenhouse gas considered as the second largest amount after carbon dioxide. Methane is released during the production and transport of coal, natural gas, and petroleum. In addition, methane is also produced by the farms and from the decay of organic waste in landfills [1].

IntechOpen

Methane is a principal component of natural gas that has the potential to be converted into valuable oxygenated products such as formaldehyde, formic acid, and especially methanol [2–4]. Methanol is one of bulk chemical that has numerous applications in chemical industries. It is used as solvent, gasoline additive, or chemical feedstock for the production of biodiesel and hundreds of other chemicals [5]. Thus, direct oxidation of methane to methanol has gained interest since conventional methods consume more energy as high temperature or pressure is required [6].

Many heterogeneous catalysts have been investigated for direct selective oxidation of methane to methanol. The main types of metal catalysts active for methane oxidation to C1-oxygenates have a degree of oxidation exceeding $^{3+}$. They include Pd, Mn, Co, Fe, V, Mo, and Ga [7]. Fe/ZSM-5 has been reported to be active for this conversion with the use of expensive N2O as oxidant [4, 8, 9]. Therefore, the use of O2 or air as oxidant is more attractive. In 2005, selective oxidation of methane into methanol using Cu/ZSM-5 was reported [10]. In 2010, Co/ZSM-5, prepared from cobalt impregnation on alkali-treated ZSM-5 zeolite aggregates, was reported to have catalytic activity in partial oxidation of methane, and selectivity and activity of catalyst depend on cobalt speciation [11].

Since 2012, our research group has conducted work on partial oxidation of methane to methanol using Co/ZSM-5 zeolite as heterogeneous catalyst. Three types of Co/ZSM-5 zeolite catalysts, i.e., cobalt-impregnated microporous ZSM-5, cobalt-impregnated hierarchical ZSM-5, and cobalt ion-exchanged hierarchical ZSM-5, have been prepared, and their catalytic activities were tested [12]. It is shown that both mesoporous properties and the type of Co species play an important role in the use of Co/ZSM-5 as heterogeneous catalyst in this reaction.

It is interesting to study further on the role of hierarchical ZSM-5 as catalyst support, especially its acidity, both Brønsted and Lewis sites. Therefore, further work has been carried out to compare the activity of hierarchical Co/ZSM-5 catalysts using the as-synthesized NaZSM-5 and H-exchanged ZSM-5 (HZSM-5) as support. The work on Co/HZSM-5 h has been explained previously, including the effect of additional trace of oxygen in the gas mixture to the yield of methanol [13]. This chapter will discuss the results from the above work, followed by exploration on the future prospect of utilize natural resources as alumina and silica precursors in ZSM-5 synthesis as well as bio-methane conversion.

2. Synthesis and characterization studies on ZSM-5 zeolites

2.1 Synthesis of hierarchical NaZSM-5 and HZSM-5 zeolites

Synthesis of hierarchical NaZSM-5 zeolite was carried out following the procedure reported by Wang et al. [14] with some modification. The gel of ZSM-5 zeolite was prepared from a homogeneous mixture with molar composition of 1 Al_2O_3: 64 SiO_2: 10 $(TPA)_2O$: 3572 H_2O. The mixture was stirred and aged at 373 K for 3 h. About 1 g PDDA was added slowly to the mixture at room temperature and stirred for 24 h. Then, the mixture was transferred into a 200 mL teflon-lined autoclave for hydrothermal process at 423 K for 144 h. Afterward, the as-synthesized zeolite was dried at room temperature followed by calcination at 823 K and labeled as as-synthesized hierarchical NaZSM-5 (NaZSM-5 h).

Hierarchical HZSM-5 zeolite was prepared through multiple ammonium exchanges (1 g of NaZSM-5 in 50 mL 1 MNH_4^+ solution). Afterward, the zeolite was dried at room temperature followed by calcination at 823 K for 3 h to obtain the HZSM-5 zeolite (labeled as HZSM-5 h).

2.2 Characterization of ZSM-5 zeolites

X-ray powder diffraction patterns were obtained with a PANalytical X-Pert Pro powder diffractometer in 5–60° 2θ with Cu Kα radiation and scanning step of 0.013° s^{-1}. Scanning electron microscope images were undertaken on a FE-SEM FEI INSPECT F50 instrument. Si, Al, and Co content was determined by using atomic absorption spectroscopy technique with the use of Shimadzu AA600 instrument. Infrared spectra were collected from pressed disks of sample mounted in a high-vacuum cell allowing in situ outgassing, using a Thermo Nicolet Nexus 470 Fourier transform infrared (FTIR) instrument at 4 cm^{-1} resolution. The surface area was calculated by using the Brunauer–Emmett–Teller (BET) method based on the adsorption data in the relative pressure (p/p_o) range 0.05–0.35. The pore-size distribution was determined by using the Barrett–Joyner–Halenda (BJH) desorption model. Micropore volume was obtained by t-plot analysis. Total pore volume was obtained from the amount of nitrogen adsorbed at p/p_o = ca. 0.99. Mesopore volume was calculated by subtracting micropore volume from total volume. The water adsorption capacity of ZSM-5 zeolites was investigated in the experiments performed using a CI microbalance. The nitrogen adsorption and desorption isotherms at 77 K were measured using a QuadraSorb Station 2 version 5.13. The samples were outgassed for 3 h at 573 K before being measured.

2.2.1 X-ray diffraction (XRD)

Figure 1 shows X-ray diffraction (XRD) patterns of various hierarchical ZSM-5 samples. As-synthesized NaZSM-5 exhibited the characteristic peaks associated with MFI framework [15]. It indicates that MFI-structured zeolites were successfully synthesized. XRD pattern of HZSM-5 shows that ZSM-5 h zeolite structure was preserved after post-synthesis treatment on NaZSM-5 h through NH_4^+ exchange process followed by calcination at 823 K.

2.2.2 Scanning electron microscope

SEM image of NaZSM-5 h (**Figure 2**) shows that NaZSM-5 particles have uniform size in coffin-like shape and the surface of particles is rough. This surface damage indicating mesoporosity were introduced, after the PDDA template was removed by calcinations [14]. Since the XRD pattern for HZSM-5 h that is prepared through ammonium

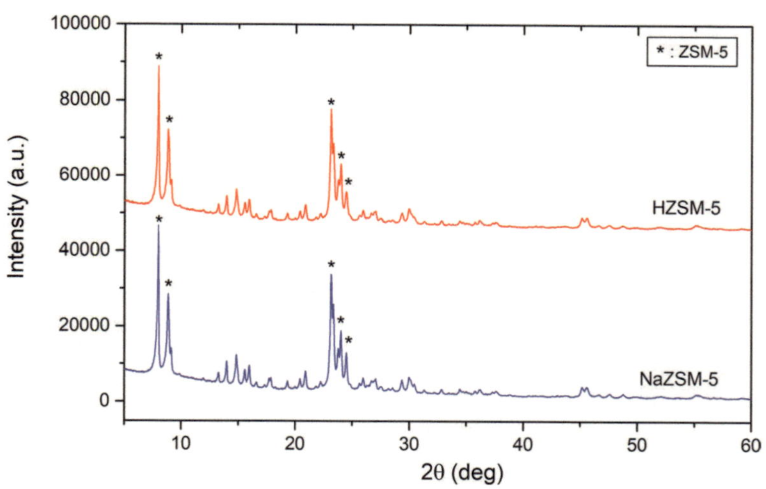

Figure 1.
XRD pattern of ZSM-5 h zeolites.

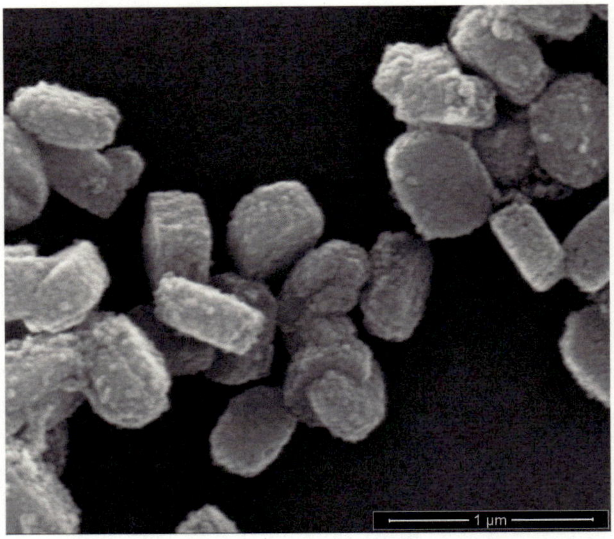

Figure 2.
SEM image of NaZSM-5 h zeolite with 100.000 magnification.

exchange and calcination at high temperature is similar to the pattern of NaZSM-5 h, it is presumed that the morphology of the crystals is also unchanged significantly.

2.2.3 Surface area analysis

Figure 3 shows N_2 adsorption and desorption isotherms of NaZSM-5 h and HZSM-5 h. The adsorption–desorption isotherms for all samples were found to belong to type IV isotherm according to BET classification, showing broad hysteresis loops indicating the presence of mesopores (ø > 2.0 nm) [16].

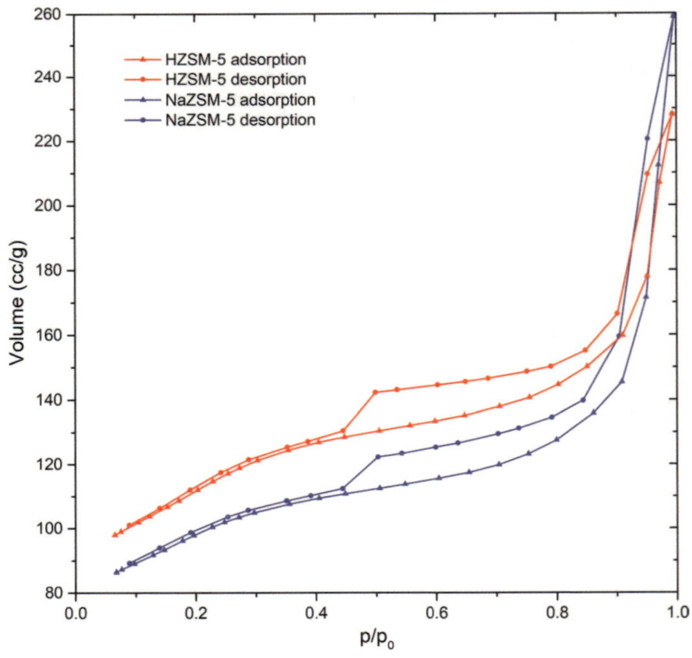

Figure 3.
Isotherm adsorption–desorption curve of ZSM-5 zeolites.

2.2.4 Physicochemical properties

Table 1 shows the physicochemical properties of the prepared ZSM-5 materials. Elemental analysis using AAS showed that Si-to-Al ratio of synthesized NaZSM-5 zeolite was 33.19. This makes ZSM-5 classified as high-silica zeolites (Si/Al ratio = ~10–100), which has played an important role as hydrocracking catalysis in petrochemistry due to its hydrophobic surfaces [17].

2.2.5 Fourier transform infrared

Figure 4 shows the infrared absorption spectra of ZSM-5 zeolites. The band at 820–650 cm^{-1} can be assigned to symmetric stretching vibration mode of

Properties	NaZSM-5	Co/NaZSM-5	HZSM-5	Co/HZSM-5
[1]Si/Al ratio	33.19	—	35.40	—
[1]Co-loading [wt.%]	—	2.485	—	2.498
[2]S_{BET} [m^2/g]	353.8	298.7	368.3	329.6
[2]V_{total}[cc/g]	0.4004	0.3213	0.3528	0.3150
[2]V_{micro}[cc/g]	0.1099	0.1020	0.1183	0.0990
[2]V_{meso}[cc/g]	0.2905	0.2193	0.2345	0.2160
[2]Pore radius [nm]	1.907	1.918	1.899	1.914
[3]% (w/w) adsorbed H_2O	17.97	—	27.78	—

[1]AAS, [2]BET, [3]microbalance experiment.

Table 1.
Physicochemical properties of ZSM-5 zeolites.

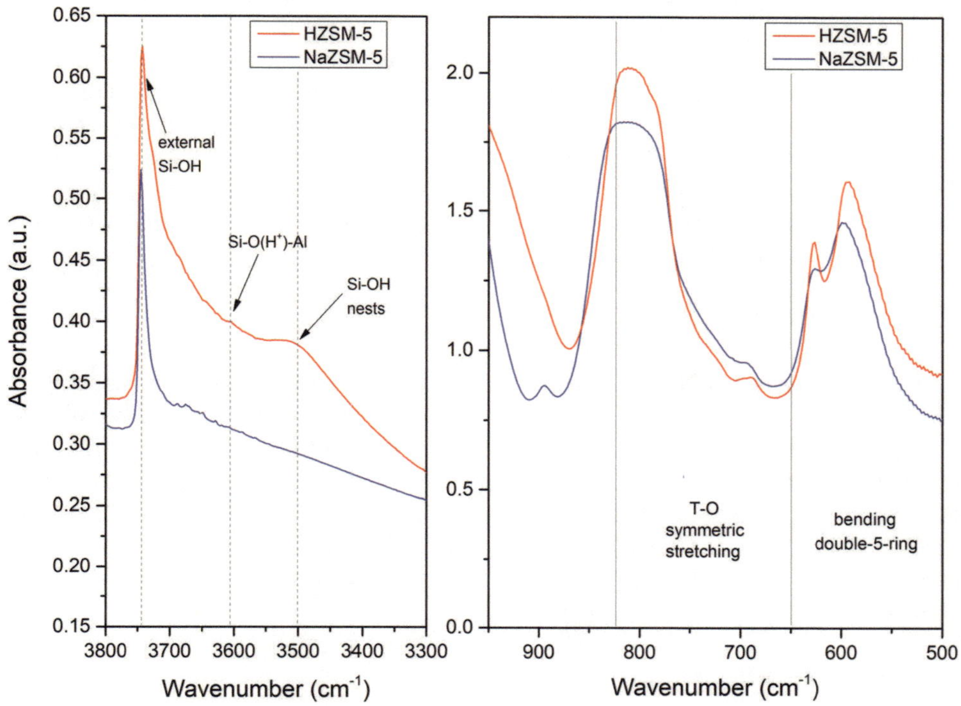

Figure 4.
FTIR absorption spectra of ZSM-5 h zeolites.

T – O (T = Si or Al) and the band at 650–500 cm^{-1} that can be attributed to the pentasil double-5-ring bending vibration [18]. Absorption bands at 3745, 3605, and 3500 cm^{-1} were observed in the infrared spectra of ZSM-5 zeolites, which are assigned for the stretching vibration of terminal Si-OH, Si-O(H^+)-Al, and Si–OH nests, respectively [19].

2.2.6 Microbalance experiments

The water adsorption capacity of NaZSM-5 h and HZSM-5 h samples, calculated from microbalance experiments, is shown in **Table 1**. After the measurement is completed, the samples were evacuated at room temperature. It is observed that the weight of the samples is decreased, close to weight before adsorption. This indicates that at room temperature, water adsorption is a reversible process as a result of water being weakly adsorbed (physisorbed) in both NaZSM-5 h and HZSM-5 h pores. The water adsorption capacity in HZSM-5 h at room temperature is higher than NaZSM-5 h. This result is consistent with the FTIR results (**Figure 4**); that is, HZSM-5 h has more silanol groups that can interact with water than NaZSM-5 h.

3. Preparation and characterization of the Co/ZSM-5 and Co/HZSM-5 catalysts

3.1 Preparation of Co/NaZSM-5 and Co/HZSM-5 zeolite catalyst

Co/NaZSM-5 h was prepared by incipient wetness impregnation method, in which 1 g of NaZSM-5 h was treated by 1.7 mL 0.2495 M Co^{2+} solution, stirred for 24 h at room temperature, and then calcined at 823 K for 3 h. Co/HZSM-5 h was prepared in a similar manner, using HZSM-5 h as starting material.

3.2 Characterization of hierarchical Co/ZSM-5 and Co/HZSM-5 zeolites

XRD, FTIR, BET, AAS, and SEM measurements were carried out on hierarchical Co/ZSM-5 and Co/HZSM-5 zeolites using similar methods as explained in 2.2, while ^{27}Al NMR spectra were recorded with a Varian Infinity Plus 400 solid-state NMR spectrometer using 90° pulses of 5 μs, a 5 s pulse delay, and a spinning rate of 4 kHz. To determine all ^{27}Al chemical shifts, kaolin was used as secondary standard. Transmission electron microscope (TEM) images were obtained using field emission TEM (300 kV) Tecnai F30 ST.

3.2.1 X-ray diffraction (XRD)

XRD pattern of Co/NaZSM-5 h and Co/HZSM-5 h in **Figure 5** shows that the ZSM-5 zeolite structure was retained after cobalt impregnation followed by calcination at 823 K on each NaZSM-5 h and HZSM-5 h, but they showed decrease in peak intensity. It is suggested that loading of Co on ZSM-5 h zeolite has caused lowering of its crystallinity. New diffraction peaks of Co_3O_4 phase (2θ = 38.6° and 47.8°) and Co_2SiO_4 phase (2θ = 56.07°) [20] are observed in XRD patterns of Co/NaZSM-5 h and Co/HZSM-5 h. This is an evidence of the presence of cobalt oxide species in the NaZSM-5 h and HZSM-5 h, respectively.

3.2.2 Surface area analysis

Figure 6 shows a decrease in surface area and total pore volume of Co/NaZSM-5 h and Co/HZSM-5 h compared to NaZSM-5 h and HZSM-5 h. More

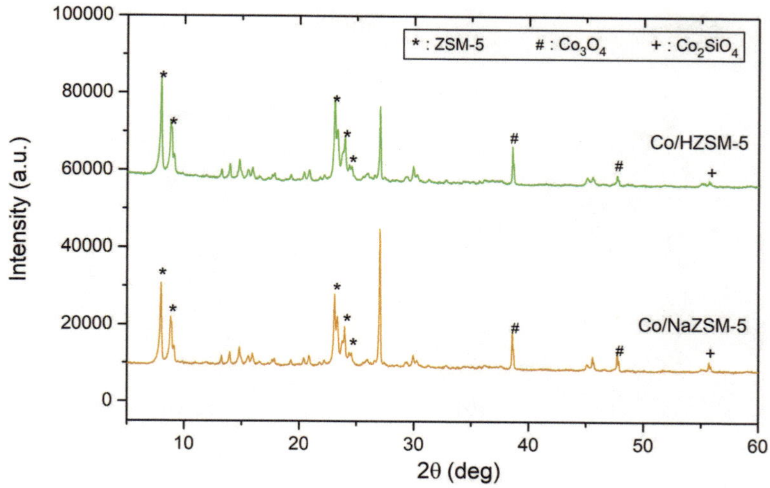

Figure 5.
XRD pattern of Co oxide-modified ZSM-5 h zeolites.

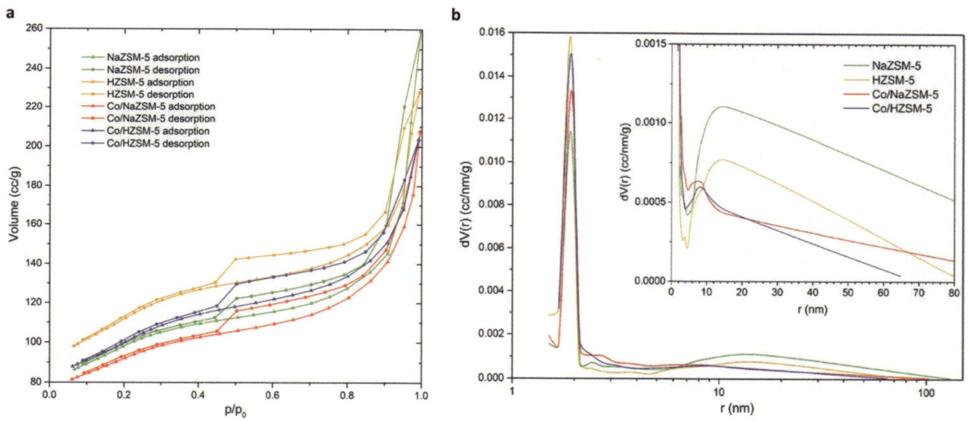

Figure 6.
(a) Isotherm adsorption–desorption and (b) pore-size distribution of Co/ZSM-5 h zeolites.

detailed information is summarized in **Table 1**. This proves that the cobalt species has penetrated into the pores and surface of ZSM-5 structure. Decrease of the mesopore volume was more significant than the micropore volume. It is assumed that due to less available spaces in micropore as they were already occupied by Na + ions, the counter ions, the Co^{2+} resides in the mesopore area. This reduction was more notable in NaZSM-5 h than HZSM-5 h, which is in accordance with the amount of Co loading shown in **Table 1**. It is suggested that zeolite acidity affects the amount of impregnated cobalt as the lower the acidity of zeolite, the higher the Co loading in zeolite.

3.2.3 TEM

Figure 7 shows the transmission electron microscope images from NaZSM-5 h to Co/NaZSM-5 h. It can be seen that the surface of NaZSM-5 consists of pores (lighter spots) that is in accordance with the rough morphology on the crystal surface observed by SEM. After being modified by cobalt oxide in Co/NaZSM-5 h, some area on the surface of the crystal became darker. This is indicative of the dispersion

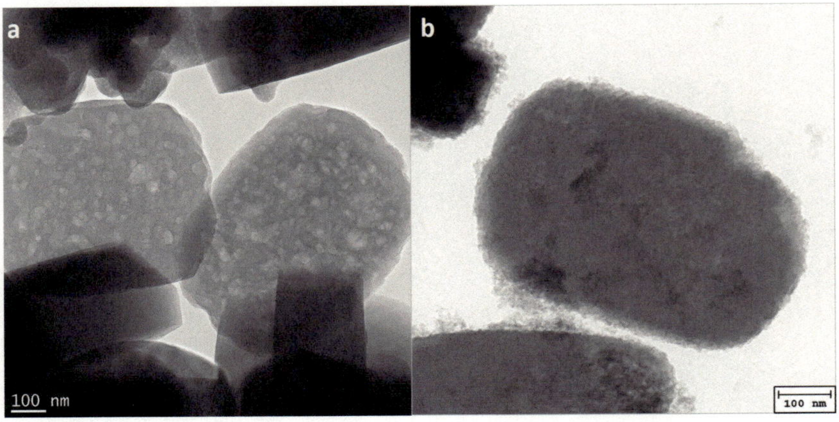

Figure 7.
TEM images of (a) NaZSM-5 h and (b) Co/NaZSM5-h.

of Co species into the pores on the surface and not creating a pool or island of cobalt oxide on the surface. This also supports the results on BET measurement that cobalt oxide prefers to reside in the mesopore.

3.2.4 Fourier transform infrared

Figure 8 shows the infrared spectra of Co/NaZSM-5 h and Co/HZSM-5 h. It can be observed that by comparing their spectra to the parent NaZSM-5 h and HZSM-5 h, respectively, it was found that cobalt impregnation decreased silanol absorption band intensity in the Si-O(H^+)-Al and Si-OH nests but not the terminal Si-OH, as deposited cobalt was covering silanol groups on the surfaces.

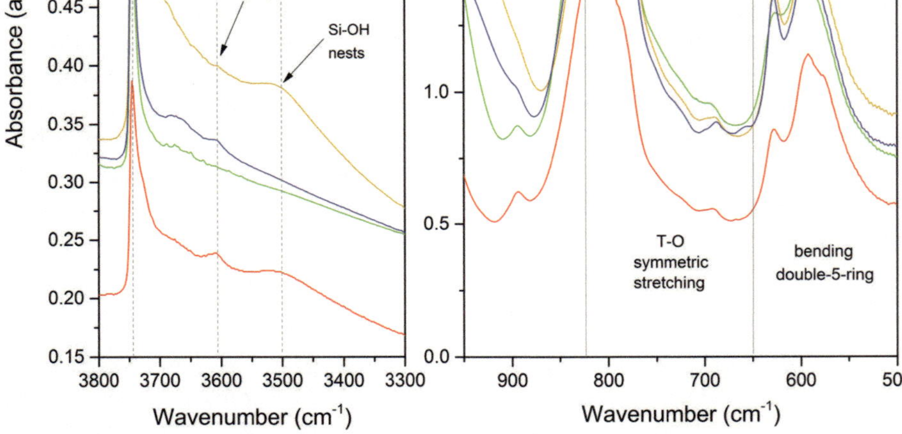

Figure 8.
FTIR absorption spectra of ZSM-5 h and Co/ZSM5-h zeolites.

3.2.5 ^{27}Al MAS NMR

Figure 9 shows ^{27}Al solid-state NMR spectra for ZSM-5 h before and after modification with acid and metal oxide. There is one main peak at 52 ppm in all ZSM-5 h zeolites, which can be assigned to the tetrahedral framework Al [16, 18]. Furthermore, ^{27}Al NMR spectra of HZSM-5 h, Co/NaZSM-5 h, and Co/HZSM-5 h were more noisy than the parent NaZSM-5 h, as a result of minor damages on zeolite framework due to acidity modification and cobalt impregnation. Furthermore, another peak at vicinity near 0 ppm, that is, attributed to octahedral non-framework Al species [21, 22], also appeared in ^{27}Al NMR spectra of Co/NaZSM-5 h and Co/HZSM-5 h.

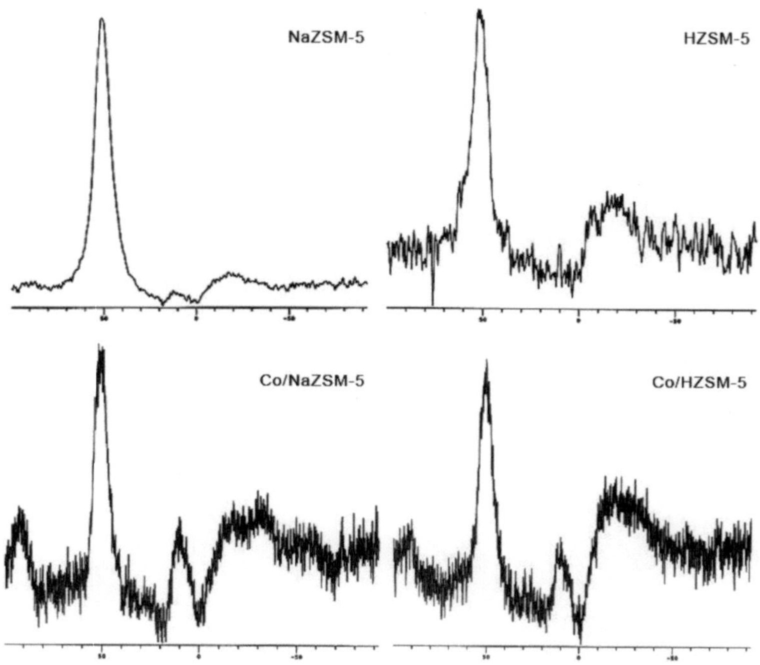

Figure 9.
^{27}Al NMR spectra of ZSM-5 h and Co/ZSM-5 h zeolites.

4. Application for partial oxidation of methane to methanol

Catalytic test on partial oxidation of methane to methanol was carried out in a stainless steel vessel as a batch reactor, following the procedure reported previously [8]. The reactor schematic diagram is shown in **Figure 10**. The catalyst used for each reaction was 0.5 g, which was activated at 773 K prior to its use. The reaction was performed at 423 K with methane pressure of 0.75 bar and a nitrogen pressure of 2 bar with a variety of reaction time of 30 minutes and 60 minutes. After the vessel was cooled down to room temperature, the product was then extracted with ethanol and characterized with GC-FID using Carbowax column. Typical chromatogram of the extracted product is shown in **Figure 11**. Methanol was determined and quantized using additional standard method. In order to observe the effect of additional O2 gas to the reaction, similar experiment was carried out using gas mixture of methane: oxygen: nitrogen (0.75: 0.1: 1.9). After the amount of methanol produced was determined, the yield of methanol was calculated as follows:

$$\text{\% Methanol Yield} = \frac{\text{mole methanol in reaction mixture}}{\text{mole methane in the input}} \times 100\% \qquad (1)$$

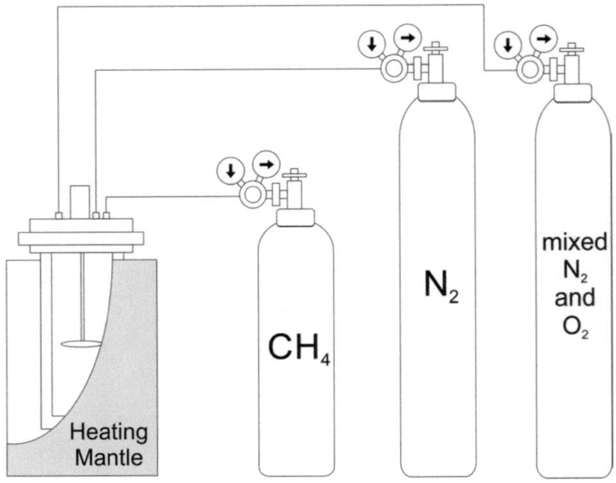

Figure 10.
Schematic for reactor of partial oxidation of methane.

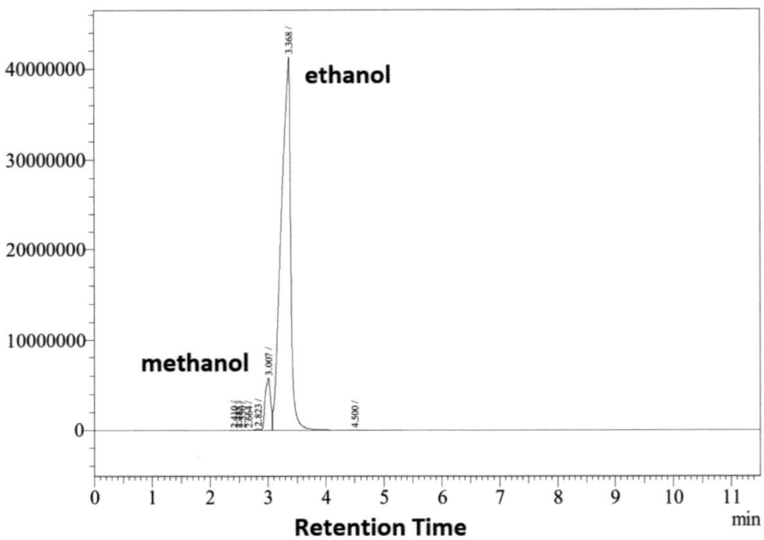

Figure 11.
Example of chromatogram from the product mixture. [Reaction condition, 0.75 bar CH4, 2 bar N2 contains 0.5% O2; catalyst, 0.5 g Co/NaZSM-5 h; T, 150°C; t, 30 minutes].

4.1 Activity of ZSM-5 h catalysts without gas O2

The first work carried out in this partial oxidation of methane to methanol was carried out using Co/NaZSM-5 h catalysts without the presence of gas O_2 [12], based on work reported in [11]. It is shown that the partial oxidation of methane to methanol could occur without the presence of molecular oxygen, suggested that the superoxide (O^{2-}) from cobalt oxide (Co_3O_4) or surface of ZSM5 framework could act as oxidation agent at 423 K. This phenomenon has been also studied computationally in [23]. **Table 2** summarizes the results in the reaction of various ZSM-5 h without additional molecular oxygen. It can be seen that using any ZSM-5 h catalysts, the percentage (%) yield of methanol increased when the reaction time was longer and Co/HZSM-5 h is more

Reaction time (minute)	Percentage (%) yield of methanol		
	NaZSM-5 h	Co/NaZSM-5 h[5]	Co/HZSM-5 h[6]
30	3.10	8.935	N/A
60	8.40	42.56	51.55

[4]Reaction condition: 0.75 bar CH4, 2 bar N2, 423 K, 0.5 g catalyst, [5][12], [6][13].

Table 2.
Percentage (%) yield of methanol from partial oxidation of methane[4].

Reaction time (minute)	Percentage (%) yield of methanol	
	Co/NaZSM-5 h	Co/HZSM-5 h[6]
30	42.00	33.22
60	N/A	79.27
120	3.70	48.70

[7]Reaction condition: 0.75 bar CH4, 2 bar N2 (with trace of 0.5% O2), 423 K, 0.5 g catalyst, [6][13].

Table 3.
Conversion of methane to methanol in the presence of oxygen[7].

active than Co/NaZSM-5 h catalyst. This could be indicative of contribution of Bronsted sites (-Si-OH, silanol groups) existing more significantly in HZSM-5 h than in NaZSM-5 h, as shown by FTIR spectra. The silanol groups would interact with produced methanol through hydrogen bond.

Furthermore, the effect of certain amount of molecular oxygen in the reaction was also studied. ZSM-5 h modified with Co oxide is able to catalyze the partial oxidation of methane with O2 as the oxidant [24]. The reaction results were tabulated in **Table 3**. It shows that the presence of molecular oxygen in the reaction system could increase the yield of methanol, so that the reaction could occur in shorter reaction time. For example, when Co/NaZSM-5 h was used in reaction time of 30 min, the percentage (%) yield of methanol for condition with additional oxygen is 42%, which is 4.7 times more than the reaction without oxygen. Similar results were also obtained using Co/HZSM-5 h as catalyst in 60 minute reaction time; the percentage (%) yield of methanol is 1.5 times higher in the presence of molecular oxygen. The effect of oxygen is more apparent when using Co/NaZSM-5 in 30 minute reaction time. Moreover, using Co/HZSM-5 h as catalyst and in the presence of oxygen, the percentage (%) yield of methanol reached the highest when the reaction took place for 60 min and then decreased after 120 minutes of reaction. The percentage (%) yield of methanol even significantly decreased when Co/NaZSM-5 h was employed for 120 min. It is suggested that the additional oxygen, in short reaction time, could regenerate the Co oxide species after their superoxides (O^{2-}) are being used to oxidize methane. However, in longer reaction time, the additional oxygen could also oxidize further the produced methanol to CO_2. As a result, complete oxidation takes place.

4.2 Reaction mechanism

Plausible reaction mechanism of partial oxidation of methane to methanol is proposed as shown in **Figure 12**. It is a modification from the mechanism reported in [23].

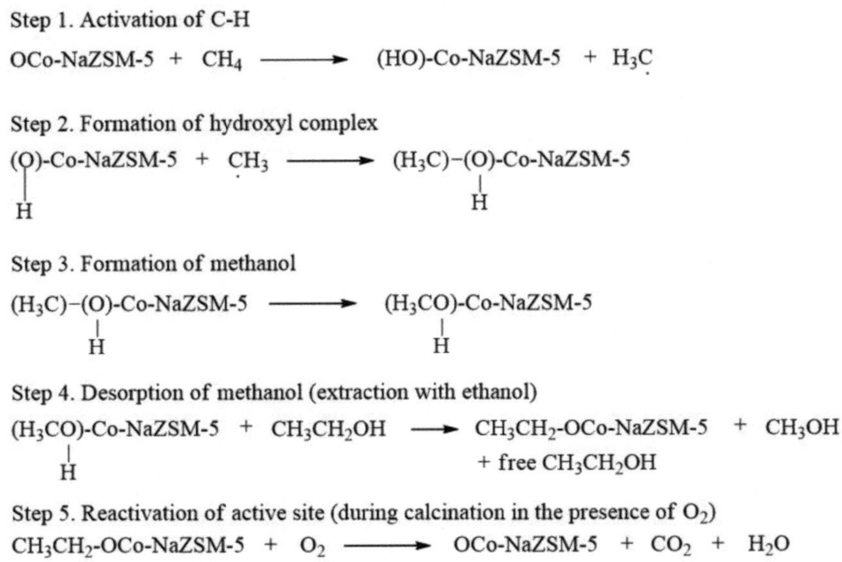

Figure 12.
Schematic diagram of plausible mechanism reaction of partial oxidation of methane on Co/NaZSM-5.

5. Future prospect

5.1 Utilization of synthesis of hierarchical NaZSM-5

Synthesis of zeolites using natural resources such as kaolin, natural zeolites, coal fly ash, and rice husk has drawn many researchers since decades ago [25]. Especially in developing countries, this work by Chareonpanich et al. [26] has been a success to the synthesis of ZSM-5 zeolite from lignite coal fly ash and rice husk to produce microporous ZSM-5. Our group has reported the synthesis of hierarchical ZSM-5 using similar procedure, with addition of secondary template (labeled as nZSM-5) [27] as the synthesis mentioned earlier in this chapter. The zeolite from this work has MFI structure although the crystallinity was lower than the pro-analysis sourced-synthetic ZSM-5. Nevertheless, the activity of Co/nZSM-5 was comparable to that of Co/cZSM-5 (prepared from commercially available NaZSM-5 with Si/Al 6.82). Synthesis of ZSM-5 using kaolin and natural zeolite and its activity as catalyst for methane oxidation is also carried out, and it is still an undergoing work.

5.2 Bio-methane as feed

Biogas is one of environmentally friendly renewable energy sources, and its utilization is very economically profitable [28]. Biogas is generally produced from anaerobic organic waste derived from the environment such as livestock manure, containing methane and carbon dioxide as the largest component and a small amount of nitrogen compounds, oxygen, hydrogen sulfide, halogen compounds, and aromatics [29]. In general, biogas can be used in combustion engines [28]; however, direct use of biogas on combustion engines can produce toxic emissions with low efficiency of biogas utilization. In our work [29], utilization of biogas as feed to be partially oxidized to methanol has been studied. Since biogas, obtained from the local biogas farm industry, consists mainly of methane (83.7%) and oxygen (16%), it is directly used without purification. The challenge faced was to compress the biogas in order to feed the reactor as much as the ultrahigh purity (UHP) methane gas. This is due to the relatively high oxygen content. Thus, the concentration of biogas (0.17 bar) fed to the reactor was much lesser than the previous UHP methane gas (0.75 bar). Nevertheless, the partial oxidation reaction of methane without additional oxygen gave 10.99% of percentage (%) yield of methanol compared to 40.56% when UHP methane was used.

6. Conclusion

Here are some remarks that could be drawn from this chapter:

1. NaZSM-5 h, HZSM-5 h, Co/NaZSM-5 h, and Co/HZSM-5 h zeolites have been characterized extensively. NaZSM-5 that is converted to HZSM-5 has more silanol functional groups and Bronsted acid sites. After impregnation with cobalt oxides, NaZSM-5 shows more decreasing pore volume and surface area than HZSM-5.

2. Acidity of ZSM-5 zeolite as support catalyst influences the percentage (%) yield of methanol, and further oxidation of methanol is decreased when acidity of zeolite is higher.

3. There is a prospect for hierarchical ZSM-5 synthesized from natural source and waste to be utilized as catalyst or support catalyst, as for partial oxidation of methane.

4. There is room for research on utilization of biogas to be partially oxidized to methanol. This also leads to other heterogeneous catalytic reactions of methane.

Acknowledgements

This work was funded by the International Collaboration Research grant from UI No. 1832/H2.R12/HKP.05.00.Perjanjian/2014, Higher Education Research grant (PUPT) from the Ministry of Research and Higher Education No. 0545/UN2.R12/HKP.05.00/2015, and PITA grant No. 2253/UN2.R3.1/KKP/05.00/2018 from UI.

Conflict of interest

There are no conflicts of interest to declare.

Author details

Yuni Krisyuningsih Krisnandi[1*], Dita Arifa Nurani[1], Muhammad Reza[1], Bayu Adi Samodro[1], Suwardiyanto[2], Nirwan Susianto[1], Arbhyando T. Putrananda[1], Indah R. Saragi[1], Aminah Umar[1,3], Sung-Min Choi[3] and Russell F. Howe[4]

1 Faculty of Mathematics and Natural Science, Department of Chemistry, Universitas Indonesia, Depok, Indonesia

2 Department Chemistry, Universitas Jember, Jember, Indonesia

3 Department of Nuclear and Quantum Engineering, Korea Advanced Institute of Science and Technology, Daejeon, Korea

4 Department of Chemistry, University of Aberdeen, Aberdeen, UK

*Address all correspondence to: yuni.krisnandi@sci.ui.ac.id

References

[1] Rotz CA. Symposium review: Modeling greenhouse gas emissions from dairy farms 1. Journal of Dairy Science. 2018;**101**(7):6675-6690. DOI: 10.3168/jds.2017-13272

[2] Arutyunov V. Direct methane to methanol: Reaction products and effect of gas composition. [Internet]. Methanol. Elsevier B.V.; 2018;**1**:173-210. DOI: 10.1016/B978-0-444-63903-5.00007-8

[3] Wang B, Albarracín-suazo S, Pagán-torres Y, Nikolla E. Advances in methane conversion processes. Catalysis Today [Internet]. 2017;**285**:1-12. DOI: 10.1016/j.cattod.2017.01.023

[4] Al-shihri S, Richard CJ, Al-megren H, Chadwick D. Insights into the direct selective oxidation of methane to methanol over ZSM-5 zeolites in aqueous hydrogen peroxide. Catalysis Today. 2017;**2018**:0-1. DOI: 10.1016/j.cattod.2018.03.031

[5] Verhelst S, Wg J, Sileghem L, Vancoillie J. Methanol as a fuel for internal combustion engines. Progress in Energy and Combustion Science. 2019;**70**:43-88. DOI: 10.1016/j.pecs.2018.10.001

[6] Han B, Yang Y, Xu Y, Etim UJ, Qiao K, Xu B, et al. A review of the direct oxidation of methane to methanol. Chinese Journal of Catalysis. 2016;**37**(8):1206-1215. DOI: 10.1016/S1872-2067(15)61097-X

[7] Mansouri S, Benlounes O, Rabia C, Thouvenot R, Bettahar MM, Hocine S. Partial oxidation of methane over Keggin-type polyoxotungstates. Journal of Molecular Catalysis A: Chemical [Internet]. 2013;**379**:255-262. DOI: 10.1016/j.molcata.2013.08.006

[8] Zuo H, Xin Q, Meynen V, Klemm E, Applied Catalysis A. General sensitivity of the selective oxidation of methane over Fe/ZSM-5 zeolites in a micro fixed-bed reactor for the catalyst preparation method. Applied Catalysis A: General. 2018;**566**(August):96-103. DOI: 10.1016/j.apcata.2018.08.022

[9] Ayodele OB. Structure and reactivity of ZSM-5 supported oxalate ligand functionalized nano-Fe catalyst for low temperature direct methane conversion to methanol. Energy Conversion and Management. 2016;**126**:537-547. DOI: 10.1016/j.enconman.2016.08.016

[10] Smeets PJ, Groothaert MH, Schoonheydt RA. Cu based zeolites: A UV – Vis study of the active site in the selective methane oxidation at low temperatures. Catalysis Today. 2005;**110**:303-309. DOI: 10.1016/j.cattod.2005.09.028

[11] Beznis NV, Van LANC, Weckhuysen BM, Bitter JH. Oxidation of methane to methanol and formaldehyde over Co – ZSM-5 molecular sieves: Tuning the reactivity and selectivity by alkaline and acid treatments of the zeolite ZSM-5 agglomerates. Microporous and Mesoporous Materials. 2011;**138**(1-3):176-183. DOI: 10.1016/j.micromeso.2010.09.009

[12] Krisnandi YK, Ary B, Putra P, Bahtiar M, Abdullah I, Francis R. Partial oxidation of methane to methanol over heterogeneous catalyst Co/ZSM-5. Procedia Chemistry. 2015;**14**:508-515. DOI: 10.1016/j.proche.2015.03.068

[13] Krisnandi YK, Samodro BA, Sihombing R, Howe RF. Direct synthesis of methanol BY partial oxidation of methane with oxygen over cobalt modified Mesoporous H-ZSM-5 catalyst. Indonesian Journal of Chemistry. 2015;**15**(3):263-268. DOI: 10.22146/ijc.1002

[14] Wang L, Zhang Z, Yin C, Shan Z, Xiao F. Hierarchical mesoporous

zeolites with controllable mesoporosity templated from cationic polymers. Microporous and Mesoporous Materials. 2010;**131**(1-3):58-67. DOI: 10.1016/j.micromeso.2009.12.001

[15] Treacy MMJ, Higgins J. Collection of Simulated XRD Powder Patterns for Zeolites. Netherland: Elsevier; 2001. DOI: 10.1016/B978-0-444-53067-7.X5470-7

[16] Thommes M, Kaneko K, Neimark AV, Olivier JP, Rodriguez-reinoso F, Rouquerol J, et al. Physisorption of gases, with special reference to the evaluation of surface area and pore size distribution (IUPAC technical report). 2015;**87**: 1051-1069. DOI: 10.1515/ci-2016-0119

[17] Li Y, Sun H, Feng R, Wang Y, Subhan F. Synthesis of ZSM-5 zeolite from diatomite for fluid catalytic cracking (FCC) application. Applied Petrochemical Research. 2015;**5**(4):347-353. DOI: 10.1007/s13203-015-0113-2

[18] Mozgawa W, Król M, Barczyk K, Science M. FT-IR studies of zeolites from different structural groups. CHEMIK. 2011;**65**(7):667-674. DOI: researchgate.net/publication/281213262%0AFT-IR

[19] Mowla O, Kennedy E, Stockenhuber M. In-situ FTIR study on the mechanism of both steps of zeolite-catalysed hydroesterification reaction in the context of biodiesel manufacturing. Fuel. 2018;**232**(January):12-26. DOI: 10.1016/j.fuel.2018.05.096

[20] El-bahy ZM, Mohamed MM, Zidan FI, Thabet MS. Photo-degradation of acid green dye over Co – ZSM-5 catalysts prepared by incipient wetness impregnation technique. Journal of Hazardous Materials. 2008;**153**:364-371. DOI: 10.1016/j.jhazmat.2007.08.060

[21] Zhang W, Han X, Liu X, Bao X. The stability of nanosized HZSM-5 zeolite: A high-resolution solid-state NMR study. Microporous and Mesoporous Materials. 2001;**50**:13-23. DOI: 10.1016/S1387-1811(01)00428-0

[22] Zhang W, Han X, Liu X, Bao X. Characterization of the acid sites in dealuminated nanosized HZSM-5 zeolite with the probe molecule trimethylphosphine. Journal of Molecular Catalysis A: Chemical. 2003;**194**:107-113. DOI: 10.1016/S1381-1169(02)00466-1

[23] Fellah MF, Onal I. Direct methane oxidation to methanol by N_2O on Fe- and Co-ZSM-5 clusters with and without water: A density functional theory study. The Journal of Physical Chemistry [Internet]. 2010;**144**: 3042-3051. DOI: 10.1021/jp9097292

[24] Bitter JH. Partial oxidation of methane over Co-ZSM-5: Tuning the oxygenate selectivity by altering the preparation route. Catalysis Letters. [Internet]. 2010;**136**:52-56. DOI: 10.1007/s10562-009-0206-6

[25] Abdullahi T, Harun Z, Othman MHD. A review on sustainable synthesis of zeolite from kaolinite resources via hydrothermal process. Advanced Powder Technology. 2017;**28**(8):1827-1840. DOI: 10.1016/j.apt.2017.04.028

[26] Chareonpanich M, Namto T. Synthesis of ZSM-5 zeolite from lignite fly ash and rice husk ash. Fuel Processing Technology. 2004;**85**:1623-1634. DOI: 10.1016/j.fuproc.2003.10.026

[27] Krisnandi YK, Yanti F, Murti S. Synthesis of ZSM-5 zeolite from coal fly ash and rice husk: Characterization and application for partial oxidation of methane to methanol series: Materials synthesis of ZSM-5 zeolite from coal fly ash and rice husk: Characterization and application for. IOP Conference Series: Materials Science and Engineering. 2017;**188**:012031. DOI: 10.1088/1742-6596/755/1/011001

[28] De AA. Biogas—A review of Sri Lanka's performance with a renewable energy technology. Energy for Sustainable Development. 2002;**6**(1):30-37. DOI: 10.1016/S0973-0826(08)60296-3

[29] Lau CS, Allen D, Tsolakis A, Golunski SE, Wyszynski ML. Biogas upgrade to syngas through thermochemical recovery using exhaust gas reforming. Biomass and Bioenergy. 2012;**40**:86-95. DOI: 10.1016/j.biombioe.2012.02.004

Section 2

Integrated Approaches of Biogas Production

Chapter 8

Contribution of Anaerobic Digestion Coupled with Algal System towards Zero Waste

Lakshmi Machineni, R. Aparna Rao and Anupoju Gangagni Rao

Abstract

Global environmental protection is of immediate concern and it can only be achieved by avoiding the use of fossil fuels. In addition, waste disposal and management could be made remunerative through the generation of renewable energy so that sustainable development is ensured. India is an agriculture-based country, and paddy residues such as rice straw and rice husk are the largest agricultural wastes in India. Currently, the common practice to dispose paddy residues is through field burning, but this has adverse effects on the air quality and consequently on people's health. However, utilization of lignocellulosic and non-food agricultural residues such as paddy residue for biogas generation by solid-stated anaerobic digestion (AD) is promising and this can substitute fossil fuels. Paddy residues for biogas production via AD has not been widely adopted because of its complex cell wall structure making it resistant to digestion by microbial attack. In addition, sequestration of carbon dioxide from biogas by algal biomass cultivated in an integrated algal bioreactor could be a promising option for biogas enrichment due to its unmatched advantages. This chapter presents the overview on utilization of non-edible residues for biogas production and its enrichment via algal biomass by means of circular bioeconomy.

Keywords: biogas, non-edible residues, algae, enrichment, zero waste

1. Introduction

Agriculture is a potential sector of the global economy, and paddy production has risen nearly 4 times in the last 60 years. India is the second largest paddy producer after China in the world, making more than 10% of the global share and thereby non-edible paddy residues, namely rice straw and rice husk [1]. The farmers are burning million of tons of paddy residues that create smog that contributes to heavy pollution in India. Furthermore, it also reduces the nutrients in the crop soil and damages desired microbial populations [1]. Therefore, management of the large quantities of paddy residues is becoming a significant environmental issue. This and the global warming effect have emerged as a research interest to utilize paddy residues as a feedstock for renewable energy like biogas production. Thus, biogas generation from paddy residues can be a major step toward harnessing one of the world's most prevalent, yet fully unutilized, renewable energy resource.

Carbon dioxide (CO_2) is a key greenhouse gas mainly releases from burning of fossil fuels namely oil, coal, and lignocellulosic biomass [2–4]. Although conventional physio-chemical and thermal approaches to reduce CO_2 have their own advantages, there is an urge to focus on biological transformation of CO_2 to energy and products of interest [5, 6]. Algae use CO_2 as a carbon source during photosynthesis and release oxygen into the atmosphere [7, 8]. In parallel to CO_2 mitigation, algal biomass has applications in production of single cell protein, bioactive compounds, pigments, cosmetics, pharmaceuticals, biogas, biofertilizer, bioplastics, biohydrogen and phytoremediation [6, 9–11]. Thus, application of micro algal biorefinery concept to produce renewable energy will enhance the economics of bioenergy production by means of circular bio-economy. The current chapter focusses on overview on anaerobic digestion, different conventional methods and microalgal biorefinery for the enrichment of biogas. In addition, the last section of the chapter discusses in short about the algal biomass recovery and its potential applications by means of circular bioeconomy.

2. Anaerobic digestion of lignocellulosic biomass

Over the past decade, anaerobic digestion (AD) has been used effectively for the degradation of agricultural lignocellulosic biomass-maize straw and wheat straw-for the production of biogas, which could be used for combined heat and power (CHP) application [12–14]. Paddy residues, composed of lignocellulose, are difficult for anaerobic microorganisms to degrade as they have a complex polymeric carbohydrates that must be preprocessed into simpler monomers-called platform molecules-that can be further converted into bioenergy [15]. A number of researchers, however, have exclusively paid attention using rice straw for biogas generation through AD [14–17], but AD of pretreated paddy residues has rarely been reported.

Controlled delignification and depolymerization of paddy residues into simpler monomers, called platform molecules, are rather challenging and specifically mandatory on a technical scale and this problem is yet to be solved, for the synthesis of bioenergy. A variety of pretreatment methods have been applied for lignocellulose biomass [14, 18, 19]. It is worth noting that the pretreatment step not only helps to release platform molecules for higher degradation by anaerobic consortia but also helps to remove toxic metal elements from biomass, which are not biodegradable and hence long-term accumulation in anaerobic digesters inhibits stable digestion of biomass in the long run. In addition, a number of important limitations such as characteristics of the pretreated digestate, different solid loadings and carbon-to-nitrogen (C/N) ratio to improve methane yield have to be investigated on immediate necessity base.

Several previous studies have reported that biogas produced from untreated rice straw is composed of methane (CH_4, ~50–75%), carbon dioxide (CO_2, ~25–50%), other impurities in small quantities such as water (H_2O, ~5–10%), hydrogen sulfide (H_2S, ~0.005–2%), siloxanes (0–0.02%), oxygen (O_2, ~0–1%) and nitrogen (0–2%) [20–23]. Biogas is enriched by removing unwanted gases (CO_2, H_2S and water vapor) to increase the calorific value, so that it is economical to compress and transport to longer places for distribution or move to other area for multifaceted applications [24–26]. Biogas production by AD is an established technology that allows farmers to generate more income from biomass waste and closing nutrient cycles [25, 27]. These synergies can be extended even further if microalgal cultivation is added to produce algae-based bioproducts. Products from microalgae, such as feed and feed additives, can again be used in the agricultural sector, which closes material cycles and extends the value chain for the biogas operator. Moreover, algae offer potential alternatives for multifaceted applications because of their high protein content and biomass yield, ability to be cultivate in their natural environment and zero effect on the food chain.

3. Biogas enrichment

Removal of CO_2 increases the percentage of biomethane in biogas. The processes involving CO_2 capture and storage are gaining attention as an alternative for reducing CO_2 concentration in the enrichment of biogas [22, 28, 29]. Several physiochemical [22, 30–32], biological [33, 34] and thermal methods [29, 35] for biogas enrichment process or purification of biogas have been reported [20, 29] (**Table 1**). For example, purification of biogas using calcium hydroxide ($Ca(OH)_2$) solution, thus CO_2 would be reacted with $Ca(OH)_2$ solution to form the precipitate of calcium carbonate ($CaCO_3$) has been reported [36]. Potential adsorbents such as activated carbon, silica gel, clay, alumina and zeolite have been reported for gas purification [37–40].

Although every adsorbent system and conventional methods have their own advantages, these technologies are considered as expensive and environmentally hazardous short-term solutions, as there are still concerns about the environmental sustainability of these processes. Alternatively, microalgal sequestration of CO_2 and its transformation to value-added biomass could be a potential solution due to its feasible and unmatched advantages over current approaches [40–42]. The waste product at the end of algal CO_2 sequestration is oxygen, clean air. Microalgae will utilize inorganic carbon from CO_2 into lipids under sunlight and increase the accumulation of biomass and algal oil [42]. As photosynthesis is a key process for microalgae metabolism and their growth, these systems are suitable even for regions with high temperatures and sunlight exposure. The effectiveness of CO_2 mitigation and its consumption by algae can modify according to the state of the algal physiology, gas residence times, light intensity, nutrient availability

Biogas enrichment methods

Physical	Chemical	Physiochemical and Thermal	Biological
• Carbon molecular sieves • Membrane separation • Water scrubbing	• Oxidations • Active coal scrubbing • Addition of iron chloride, calcium chloride • Chemical absorption with sodium hydroxide	• Cryogenic cooling • Gas-liquid absorption membranes • High pressure gas separation • Polyethylene glycol absorption • Solid oxide fuel cells	• Biofiltration • Suspended growth systems • Biofilm growth systems • Algal system

Table 1.
Different methodologies for biogas purification.

and temperature. It has been reported that CO_2 sequestration as high as 99% is attainable upon defined environmental and nutritional conditions, and with gas residence times as low as two seconds [42–45]. In addition, microalgal system combined with AD systems and the synergy between algae-bacteria can help to avoid the power demands from aeration, which actually represent almost 60% of the total energy requirement of waste effluent treatment plants in industries. During photosynthesis, algal system provides oxygen that is necessary for aerobic microbes to digest and biodegrade organic effluents, consuming in turn the CO_2 released due to bacterial growth [44].

4. Microalgae

Microalgae are phototrophic microorganisms that generate biomass with simple nutritional and low light energy and CO_2 requests. These are photosynthetically highly efficient (~10–20%) in comparison with terrestrial plants (1–2%) to fix CO_2. It was reported that more than 100,000 species exist. Advantages of being sustainable at high flue gas concentration and cogeneration of top-value products put these as the preferential and potential organisms (**Figure 1**). Microalgae have the ability to synthesize high amounts of proteins nearly 51–71% of dry matter compared to meat, 43%, soybeans 37%, milk 26% and rice 8%, which are essential for use in human and animal food supplements. Not only proteins, microalgae carbohydrates, ~25% of its dry matter, are made in the forms of simple mono- and polysaccharides, which are easy to digest. Algae are the best candidates for the production of biodiesel as they do not compete with edible crops and can produce up to 80,000 L of oil per acre per year, which is almost 31 times higher than biodiesel produced by the best terrestrial crop, namely palm tree. Moreover, biomass harvested and dewatered from microalgae belongs to the groups of *Spirulina*, *Chlorella*, *Dunaliella*, *Nostoc*

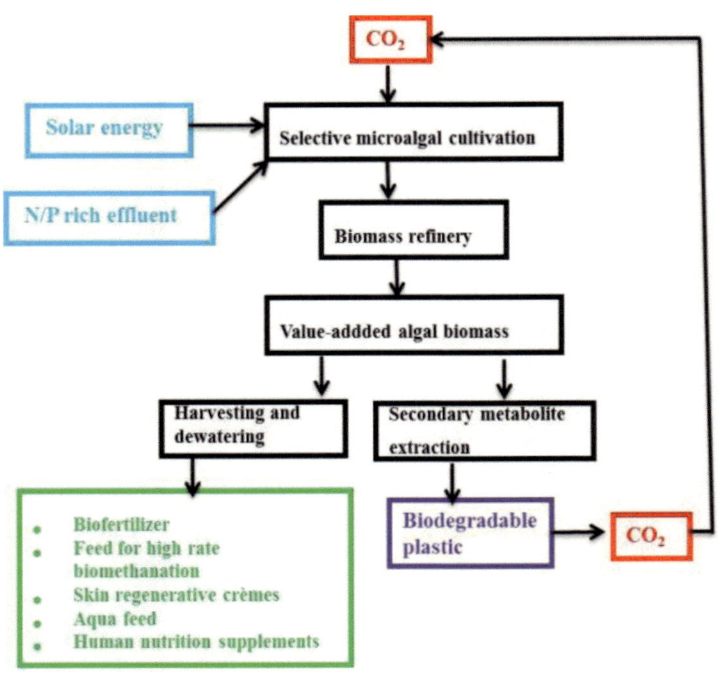

Figure 1.
Schematic algal biorefinery showing different products that can be obtained from microalgal biomass.

and *Aphanizomenon* and is available in human healthy foods industry in the form of powders, capsules, tablets, pastilles and liquids [46–48].

Some algae can increase their biomass with double growth rate within 3.5 h in their exponential phase. Algae growth yield is three to four times higher in the presence of soluble gases, namely CO_2 and H_2S [49, 50]. As the biogas passes through algal reactor, methane, which is not soluble, flashes off, whereas CO_2 and H_2S essentially infuse and completely dissolve in liquid stream. By allowing biogas stream which is typically composed of 60% methane, 39% CO_2 and less than 1% H_2S passes through coupled algal reactor, will transform to a biogas that is over 90% methane and the CO_2 and H_2S being reduced by 85–95% by algae biomass [49, 50].

Biomass cultivated in photobioreactors can be utilized for several applications, including substrate for bioenergy such as biogas, biofuels, biofertilizers, biosorbents and biopolymers [50, 51]. For instance, biopolymers recovered from algae can be adapted into packaging materials and have the advantage of being renewable [52].

What really makes the use of algae a thriving technology is that these microorganisms have the potential to efficiently remove nutrients from wastewater and provide high-value biomass energy with low cost. Enclosed bioreactors and open ponds are the two predominant methods for microalgal cultivation (**Figure 2**) [53]. Interestingly, closed photoreactors provide sterility and allow for much greater control over culture parameters such as light intensity, CO_2, nutrient levels and temperature, and thus higher biomass productivities can be reached [11]. In parallel to CO_2 mitigation, algal biomass has applications in human nutritional supplements such as vitamins, Omega-3 fatty acids, biotin, production of antiaging creams, anti-irritant creams, skin regenerate creams, biogas, biofertilizer, aqua and animal feed, and treatment of waste water [46].

Recently algae-based strategies for the removal of toxic minerals such as arsenic (As), bismuth (Bi), bromium (Br), cadmium (Cd), chromium (Cr), mercury (Hg) and lead (Pb) have also been reported individually or in a mixture, and some commercial applications have been initiated [54–56]. Therefore, a sustainable closed loop microalgae-mediated CO_2 sequestration system could be integrated with biogas generation infrastructure after optimization of algal cultivation system and key process parameters, and recovery of novel bioproducts from harvested microalgal biomass.

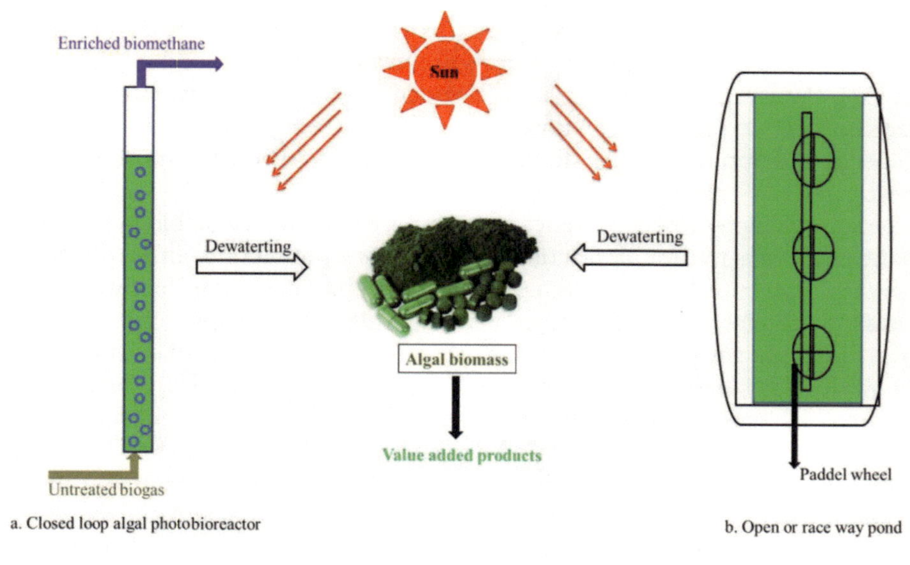

Figure 2.
Two different methods of microalgal cultivation.

5. Biomass harvesting and dewatering

Development of biorefinery for production and conversion of algal biomass will be the unified solution to meet the day-by-day increasing energy demand and to reduce risks associated with global warming due to tailpipe emissions. The microalgal-based production chain is classified into three series of steps, namely biomass inoculation and cultivation, harvesting and dewatering, and extraction of concentrated biomass for desired applications. The systems for algal cultivation can be tanks, trays, open ponds, closed or hybrid photobioreactors. It has been suggested that these systems deliver a very dilute biomass concentration ranging from 0.05 to 0.075% dry matter for open pond systems and 0.3–0.4% for closed reactors [57]. Hence, there is an immediate requirement to develop an efficient algal dewatering process to reach the biomass up to 30% in total dry product. Concentrating algal biomass and purifying it into products from broth occur in two stages: a single step of harvesting followed by one- or two-step separate biomass dewatering, which is then fractionated and extracted to extend the "shelf-life of biomass" and to make the product accessible for further application [58, 59].

Recent advance and novel high-tech research in bioprospecting new strains, breakthrough innovations in culture cultivations and complete process optimization are certainly increasing our hope about the forthcoming achievements by microalgal biorefinery. However, the potential of successful commercial deployment is associated with simple and indigenous innovations in downstream operations, specifically cell harvesting, cell disruption and extraction, which can actually cut down the costs at a biorefinery level, along with process integration. During algal biomass cultivation, the harvesting process is the main constraint, representing more than 20% of the total production cost due to low biomass density (0.2–2 $g\ L^{-1}$) and small size (10–30 μm). The methods for harvesting either used independently or in combination. However, most of these methods still involve economic or technological drawbacks, such as a high-energy cost (centrifugation), algal biomass contamination (chemical flocculation), or nonfeasibility of scaling-up. Besides the operational cost, concerning selection of the adequate harvesting method, several aspects such as the following should be considered: (i) harvesting speed, (ii) harvesting efficiency, and (iii) density and quality of biomass in the resultant concentrate [60, 61]. Among different polymers, the chitosan prepared from the waste of white shrimp is reported as a good cost-effective and efficient flocculant for algal biomass because of its properties such as faster deposition rate [57]. However, the optimal pH, chitosan dosage, chitosan physiochemical characteristics and flocculation time to achieve ~100% of algal biomass harvesting efficiency and optimization of storage condition for harvested biomass should be investigated more clearly for further applications.

Because of high protein content and biomass yield, microalgal biomass offers a potential alternative for bioplastic and biofertilizer production, either directly or in secondary metabolites form. Dewatered algal biomass can be modified into bioplastic and biofertilizer. Bio-based plastics help to "decarbonise" the economy. However, unlike soy protein isolate or feather meal protein, it is not economical or technically feasible to extract the protein from the algal biomass [11]. Consequently, more research must be developed aiming to optimize the extraction of secondary metabolites to create a sustainable and biodegradable alternative to fossil fuel–based plastics. After secondary metabolite extraction, the residual algal biomass can be reused as feed for biogas production via anaerobic digestion and biofertilizer. Several researchers have developed an indigenous assembly of macroalgae, which was installed and grown in CO_2- infused wastewater effluent (**Figure 2**) [62–64]. Later, algal biomass was harvested and co-digested with sewage sludge to enhance bio-methane production. Several techno-economic constraints have to be solved for the generation of biomethane from algal biomass is economically feasible [63, 65]. For instance, potential issues to be focussed further to enhance biomass conversion to biomethane are high sensitivity of methanogenic microorganisms, unbalanced C/N ratio of algal biomass, and high lipid contents, and cost associated with biomass recovery [65].

In addition to different solutions to huge environmental problems like deficiency of nitrogen content in the soil composting causing pollution must work in parallel with other action. Algae can serve the purpose by fixing atmospheric nitrogen and synthesizing plant growth promoters as nitrogen content of the soil is the second major factor affecting plant growth after water [66, 67]. Biofertilizers made from algae will be an effective replacement for chemical fertilizers by means of circular bioeconomy. Thus, the application of microalgal biorefinery concept to produce renewable energy will enhance the economics of bioenergy production by means of circular bioeconomy.

6. Conclusions

Lignocellulosic biomass has a huge potential for biogas production, which would be a sustainable alternative for nonrenewable fossil fuels. Upon optimization of lignocellulosic biomass delignification, high rate of biomethanation would be a possible feasible solution to fill previous research gaps without using costly fermentable sugars from food sources. Most importantly, having a clearer understanding on biomass characteristics before and after pretreatment will provide information about storage of delignified feedstock, which is key point for saving renewable biomass in order to meet energy demand of future generation before it gets exploited like fossil fuels. Later, presenting the CO_2 biocapture by selective algal strain could reduce cost associated with conventional chemical CO_2 scrubbing technologies and could be a breakthrough with potential applications. In addition, algal flocculant chitosan can be extracted from shrimp waste, which will be cost-effective for harvesting and dewatering of algal biomass by means of circular bioeconomy. It is most important to achieve ~100% harvesting of microalgal biomass with chitosan under optimal physiochemical parameters associated with flocculation approach. Polysaccharides and oils recovered from dewatered algae can adapt as lightweight, waterproof biodegradable plastic material, and organic fertilizer to enrich crop yield. However, more research on anaerobic digestion of lignocellulosic biomass coupled with algal systems could be useful to commercialize biogas production from widely produced low-cost substrate, and cultivation of algae in parallel for CO_2 mitigation results in top-value chemical generation with home-grown technology.

Author details

Lakshmi Machineni[1,2]*, R. Aparna Rao[2] and Anupoju Gangagni Rao[2]

1 CSIR-IICT, Hyderabad, India

2 Bioengineering and Environmental Sciences (BEES) Group, Department of Energy and Environmental Engineering (DEEE), CSIR-Indian Institute of Chemical Technology (IICT), Hyderabad, Telangana State, India

*Address all correspondence to: machineni.402@csiriict.in

References

[1] Bhuvaneshwari S, Hettiarachchi H, Meegoda JN. Crop residue burning in India: Policy challenges and potential solutions. International Journal of Environmental Research and Public Health. 2019;**16**(5):832

[2] Yan N et al. The potential for microalgae as bioreactors to produce pharmaceuticals. International Journal of Molecular Sciences. 2016;**17**(6)

[3] Mardani A et al. Carbon dioxide (CO_2) emissions and economic growth: A systematic review of two decades of research from 1995 to 2017. Science of The Total Environment. 2018;**649**

[4] Darkwah WK et al. Greenhouse effect: Greenhouse gases and their impact on global warming. Journal of Scientific Research and Reports. 2018;**17**:1-9

[5] Muhammad R et al. Exploring the potential of microalgae for new biotechnology applications and beyond: A review. Renewable and Sustainable Energy Reviews. 2018;**92**

[6] Sun X-M et al. Microalgae for the production of lipid and carotenoids: A review with focus on stress regulation and adaptation. Biotechnology for Biofuels. 2018;**11**(1):272

[7] Fulke A et al. CO_2 sequestration by microalgae: Advances and perspectives. 2014:1-7

[8] Cea-Barcia G et al. Biogas purification via optimal microalgae growth: A literature review. Biotechnology Progress. 2018;**34**(6):1513-1532

[9] Wollmann F et al. Microalgae wastewater treatment: Biological and technological approaches. Engineering in Life Sciences. 2019

[10] Zhao X et al. Evaluation of revolving algae biofilm reactors for nutrients and metals removal from sludge thickening supernatant in a municipal wastewater treatment facility. Water Research. 2018;**143**:467-478

[11] Sharma N, Sharma P. Industrial and biotechnological applications of algae: A review. Journal of Advances in Plant Biology. 2018;**1**

[12] Croce S et al. Anaerobic digestion of straw and corn stover: The effect of biological process optimization and pre-treatment on total bio-methane yield and energy performance. Biotechnology Advances. 2016;**34**

[13] Yong Z et al. Anaerobic co-digestion of food waste and straw for biogas production. Renewable Energy. 2015;**78**:527-530

[14] Prajapati KK, Pareek N, Vivekanand V. Pretreatment and multi-feed anaerobic co-digestion of agro-industrial residual biomass for improved biomethanation and kinetic analysis. Frontiers in Energy Research. 2018;**6**(111)

[15] Sawatdeenarunat C et al. Anaerobic digestion of lignocellulosic biomass: Challenges and opportunities. Bioresource Technology. 2015;**178**:178-186

[16] Hutnan M. Maize Silage as Substrate for Biogas Production. 2016

[17] Zieliński M et al. Effects of nutrients supplementation on enhanced biogas production from maize silage and cattle slurry mixture. Water, Air, & Soil Pollution. 2019;**230**(6):117

[18] Machineni L. Lignocellulosic biofuel production: Review of alternatives. Biomass Conversion and Biorefinery. 2019

[19] Zheng Y et al. Pretreatment of lignocellulosic biomass for enhanced biogas production. Progress in Energy and Combustion Science. 2014;**42**:35-53

[20] Awe O et al. A review of biogas utilisation, purification and upgrading technologies. Waste and Biomass Valorization. 2017;**8**

[21] Bušić A et al. Recent trends in biodiesel and biogas production. Food Technology and Biotechnology. 2018;**56**(2):152-173

[22] Al Mamun MR, Torii S. Enhancement of methane concentration by removing contaminants from biogas mixtures using combined method of absorption and adsorption. International Journal of Chemical Engineering. 2017;**2017**:1-9

[23] Kuruti K et al. Exploitation of rapid acidification phenomena of food waste in reducing the hydraulic retention time (HRT) of high rate anaerobic digester without conceding on biogas yield. Bioresource Technology. 2017;**226**:65-72

[24] Caruso MC, Braghieri A, Capece A, Napolitano F, Romano P, Galgano F, et al. Recent updates on the use of agro-food waste for biogas production. Applied Sciences. 2019;**9**:1217

[25] Lauer M et al. Making money from waste: The economic viability of producing biogas and biomethane in the Idaho dairy industry. Applied Energy. 2018;**222**:621-636

[26] Kadam R, Panwar NL. Recent advancement in biogas enrichment and its applications. Renewable and Sustainable Energy Reviews. 2017;**73**:892-903

[27] Anupoju GR et al. Biogas from poultry litter: A review on recent technological advancements. 2015. pp. 133-147

[28] Angelidaki I et al. Biogas upgrading and utilization: Current status and perspectives. Biotechnology Advances. 2018;**36**

[29] Kapdi S et al. Biogas scrubbing, compression and storage: Perspective and prospectus in Indian context. Renewable Energy. 2005;**30**:1195-1202

[30] Jha B et al. Biogas—A sustainable and potential fuel for transport application. Journal of Biofuels and Bioenergy. 2015;**1**

[31] Kohn MP, Castaldi MJ, Farrauto RJ. Biogas reforming for syngas production: The effect of methyl chloride. Applied Catalysis B: Environmental. 2014;**144**:353-361

[32] Jiang Y et al. Simultaneous biogas purification and CO_2 capture by vacuum swing adsorption using zeolite NaUSY. Chemical Engineering Journal. 2018;**334**:2593-2602

[33] Aryal N et al. An overview of microbial biogas enrichment. Bioresource Technology. 2018;**264**

[34] Wang W et al. Performance and microbial community analysis of the anaerobic reactor with coke oven gas biomethanation and in situ biogas upgrading. Bioresource Technology. 2013;**146**:234-239

[35] Zhao Q et al. Purification technologies for biogas generated by anaerobic digestion. In: Purification Technologies for Biogas Generated by Anaerobic Digestion. 2010. pp. 1-24

[36] Masyhuri AP, Ahmad AM, Djojowasito G. Rancang Bangun Sistem Penyerap Karbon dioksida (CO_2) Pada Aliran Biogas Dengan Menggunakan Larutan $Ca(OH)_2$. Jurnal Keteknikan Pertanian Tropis dan, Biosistem. 2013;**1**:19-28

[37] Montanari T et al. CO_2 separation and landfill biogas upgrading: A

comparison of 4A and 13X zeolite adsorbents. Energy. 2011;**36**(1):314-319

[38] Ricaurte Ortega D, Subrenat A. Siloxane treatment by adsorption into porous materials. Environmental Technology. 2009;**30**(10):1073-1083

[39] Paolini V et al. Biogas cleaning and upgrading with natural zeolites from tuffs. Environmental Technology. 2015;**37**:1-33

[40] Widyanuriawan D, Sugiarto. Biogas purification using natural zeolite and NaOH. Applied Mechanics and Materials. 2014;**664**:415-418

[41] Ramaraj R, Dussadee N. Biological purification processes for biogas using algae cultures: A review. International Journal of Sustainable and Green Energy. 2015;**4**:20-32

[42] Sundaramurthy S, Premalatha M. An overview of CO_2 mitigation using algae cultivation technology. International Journal of Chemical Research. 2011;**3**:110-117

[43] Bahr M et al. Microalgal-biotechnology as a platform for an integral biogas upgrading and nutrient removal from anaerobic effluents. Environmental Science & Technology. 2014;**48**(1):573-581

[44] Wang X et al. Screening of microalgae for integral biogas slurry nutrient removal and biogas upgrading by different microalgae cultivation technology. Scientific Reports. 2017;7(1):5426

[45] Sayre R. Microalgae: The potential for carbon capture. BioScience. 2010;**60**(9):722-727

[46] Khan MI, Shin JH, Kim JD. The promising future of microalgae: Current status, challenges, and optimization of a sustainable and renewable industry for biofuels, feed, and other products. Microbial Cell Factories. 2018;**17**(1):36-36

[47] Cuellar-Bermudez SP et al. Photosynthetic bioenergy utilizing CO_2: An approach on flue gases utilization for third generation biofuels. Journal of Cleaner Production. 2015;**98**:53-65

[48] Barkia I, Saari N, Manning SR. Microalgae for high-value products towards human health and nutrition. Marine Drugs. 2019;**17**(5):304

[49] Adamczyk M, Lasek J, Skawińska A. CO_2 biofixation and growth kinetics of *Chlorella vulgaris* and *Nannochloropsis gaditana*. Applied Biochemistry and Biotechnology. 2016;**179**(7):1248-1261

[50] Meier L et al. Removal of H_2S by a continuous microalgae-based photosynthetic biogas upgrading process. Process Safety and Environmental Protection. 2018;**119**:65-68

[51] Koutra E et al. Bio-based products from microalgae cultivated in digestates. Trends in Biotechnology. 2018;**36**

[52] Özçimen D et al. A review on algal biopolymers. Journal of Chemical Engineering Research Updates. 2017;**4**:7-14

[53] Granata T. Dependency of microalgal production on biomass and the relationship to yield and bioreactor scale-up for biofuels: A statistical analysis of 60+ years of algal bioreactor data. BioEnergy Research. 2017;**10**(1):267-287

[54] Gani P et al. Green approach in the bio-removal of heavy metals from wastewaters. MATEC Web of Conferences. 2017;**103**:06007

[55] Heidarpour A et al. Bio-removal of Zn from contaminated water by using green algae isolates. Environmental Technology & Innovation. 2019;**16**:100464

[56] Kaur H, Rajor A, Kaleka A. Role of phycoremediation to remove heavy metals from sewage water: Review article. Journal of Environmental Science and Technology. 2019;**12**:1-9

[57] Zhu L, Li Z, Hiltunen E. Microalgae *Chlorella vulgaris* biomass harvesting by natural flocculant: Effects on biomass sedimentation, spent medium recycling and lipid extraction. Biotechnology for Biofuels. 2018;**11**:183-183

[58] Wan C et al. Current progress and future prospect of microalgal biomass harvest using various flocculation technologies. Bioresource Technology. 2015;**184**:251-257

[59] Li Y et al. Flocculation mechanism of *Aspergillus niger* on harvesting of *Chlorella vulgaris* biomass. Algal Research. 2017;**25**:402-412

[60] Al hattab M. Microalgae harvesting methods for industrial production of biodiesel: Critical review and comparative analysis. Journal of Fundamentals of Renewable Energy and Applications. 2015;**05**

[61] Malik S et al. Microalgal flocculation: Global research progress and prospects for algal biorefinery. Biotechnology and Applied Biochemistry. 2019

[62] Bohutskyi P et al. Co-digestion of wastewater-grown filamentous algae with sewage sludge improves biomethane production and energy balance compared to thermal, chemical, or thermochemical pretreatments. Frontiers in Energy Research. 2019;7

[63] Bohutskyi P. Biogas production from algae and cyanobacteria through anaerobic digestion: A review, analysis, and research needs. 2013:873-975

[64] Wu N et al. Techno-economic analysis of biogas production from microalgae through anaerobic digestion. 2019

[65] Milledge J et al. A brief review of anaerobic digestion of algae for bioenergy. Energies. 2019;**12**:1166

[66] Chatterjee A et al. Chapter 10—Role of algae as a biofertilizer. In: Rastogi RP, Madamwar D, Pandey A, editors. Algal Green Chemistry. Amsterdam: Elsevier; 2017. pp. 189-200

[67] Pabbi S. Blue green algae: A potential biofertilizer for rice. 2015. pp. 449-465

Chapter 9

Exploitation of Digestate in a Fully Integrated Biowaste Treatment Facility: A Case Study

Maria Laura Mastellone

Abstract

The increase of biowaste generation has reached critical levels in many countries. The European legislation introduced the biowaste treatment and the organic recycling as central theme of its political agenda with the aim to promote the sustainable exploitation of this peculiar waste. The most utilized technologies applied to the biowaste treatment are based on the biological processes targeting to produce biogas or, more recently, biomethane to be used as fuel. The production of biomethane allows to produce a substitute of the fossil methane with a yield of about $0.07 g_{CH4}/g_{biowaste}$; the remaining fractions are waste coming from the pretreatment/refining steps, solid digestate or stabilized compost, and leachate. The sustainable treatment of these fractions is a mandatory issue to treat the biowaste in a reliable and sustainable integrated process since their amount is more than 85% and the impact of their treatment on environment and economy of the overall treatment process can be quite relevant. This chapter focused on the so-called smart facility that integrates processes based on thermochemical processes with the biological one targeting to increase the overall sustainability, the flexibility regarding the input biowaste composition, and the independency by the external factors affecting the waste trading.

Keywords: anaerobic digestion, biomethane, hydrochar, carbonization, gasification, smart facility, sustainable waste management

1. Introduction

Biowaste can be defined as a mixture of vegetal and animal biomasses that are residues of human activities. It contains large carbon-based macromolecules that can be used as a renewable material for energy production, carbon sequestration, and soil conditioner and fertilizer. Nevertheless, biowaste management requires a specific attention in order to increase the overall sustainability of the treatment and to define guidelines to increase its effective recovery. In Europe, more than 118 millions of tons of biodegradable waste are produced each year; only 25% of this amount is collected and sent to the given recovery option [1]. The recovery treatments are mainly based on biological processes: the most used is the composting (low capital and operating cost, low value of products) followed by the integrated anaerobic/aerobic digestion (medium–high capital cost, low operating cost, medium value of products, depending of country incentives program). Although the composting is widely used, its sustainability is not always guaranteed because of long process time, large areas needed for storage and processing, environmental impact due to annoying odors released by diffuse and fugitive emissions other than a not favorable ratio between the value of the product (compost), and the cost of the process. The recourse to an anaerobic digestion as preliminary stage allows to improve the overall process by permitting the production of biogas in addition to the compost. The main limitation of biological process is the low economic value of the compost obtained from biowaste coming from separate collection of municipal waste. This important source of biodegradable matter is often contaminated by other waste with a fraction between 10 and 25%, depending on the waste collection system adopted for the separate collection [2–4]. The presence of this fraction, generally represented by plastics and metals, can further decrease the economic value of the compost that is sold at a price between 0 and 3 €/Mg [5].

An alternative to compost production is the transformation of the biowaste, including digestate, into different products, either solid, liquid, or gaseous obtained by means of thermochemical treatments. Depending on the specific production process and feedstock, the obtained products are different: thermal decomposition of wood, peat, or some related natural organic materials produces charcoal [6, 7]; the torrefaction produces biocoal [8]; if the charred organic matter is applied to soil with the intent to improve soil properties, it is called biochar [9]; moreover, the product of hydrothermal pyrolysis, is called hydrochar [10].

The hydrothermal pyrolysis (HTC) converts all substrates containing carbohydrates and molecules, including biowaste, into hydrochar, gas, and leachate by means of extraction of nitrogen and oxygen in a subcritical water environment [11]. The HTC stage can be applied to the fresh biowaste or to the digestate produced by the anaerobic digestion plants. In this latter case, the integration is able to avoid the aerobic treatment that is time- and space-consuming and obtain a high-added value product, in a limited footprint. The other waste stream that needs to be exploited is the not biodegradable waste; the fate of this waste is the landfilling or the energy recovery by combustion [12], and the gate fee for its disposal strongly increased in the last years. The thermochemical processes applicable to this kind of waste are pyrolysis and gasification: the latter is preferable since it is energetically self-sustainable and allows to produce both heat and electricity [13–15]. The main advantage of gasification is the limited size of the plant and the possibility to install it with capacities starting from few hundreds of kilograms in an hour.

The present chapter aimed to demonstrate the feasibility of integration between small-scale thermochemical processes and the biowaste biological treatment facility with the target to reduce the waste production, increase the energy recovery, and, more in general, increase the sustainability of the plant.

2. Configuration of the base case anaerobic digestion facility

2.1 Description of the unit processes

The standard configuration of anaerobic digestion facility consists of the following sections:

a. Acceptance, weight, and discharge of biowaste from the lorries

b. Preliminary mechanical treatment and sorting

c. Mixing between substrates having different moisture content and structure (e.g., lignin-based biowaste is added to food biowaste to increase permeability, moderate moisture content, and modify the C/H ratio)

d. Pulping (only in case of wet digestion processes)

e. Anaerobic digestion into one (unique bioreactor) or two stages (two in series reactors for hydrolysis and acetogenesis/methanation steps)

f. Leachate recirculation and storage before treatment or delivery to the water treatment plant

g. Aerobic stabilization of digestate (composting)

h. Mechanical refining and foreign matter removal

The data and the information utilized in this work are based on a full-scale facility that utilizes a dry-batch technology to perform the anaerobic digestion, and it is integrated with the composting plant to obtain the mineralization of the digestate. The facility is located in the industrial area of Naples, Italy [5], and treat biowaste from household separate collection and restaurants.

The block diagram of the integrated processes is reported in **Figure 1**.

With reference to the unit processes labels reported in **Figure 1**, a short description is reported in the following paragraphs.

2.1.1 Pretreatment and sorting

The organic fraction of municipal solid waste contains a certain amount of foreign matter constituted by inorganic and organic nonbiodegradable materials such as glass, ceramic, metals, plastic bags, plastic closures, wires, etc. The size distribution of this fraction ranges from few millimeters up to several centimeters, allowing the removal of large objects by means of manual sorting and sieving in a trommel. Generally, the minimum size of the holes installed to remove the foreign matter is 5 cm. A photo of the waste removed by using the mechanical sorting after the bag opener (light-intensity shredding) is reported in **Figure 2**.

2.1.2 Anaerobic digestion

The anaerobic digestion is then carried out by using eight batch reactors operated by recurring to an operation mode by including the following phases: emptying (a), filling and mixing (b), and reaction (c). The reactors are sequentially operated in order to have a semicontinuous operation. Steps (a) and (b) require a couple of days to be carried out, that is why each bioreactor starts the reaction phase with a delay of two days; the reaction phase has a duration of 28 days. Each reactor is filled with about 200 t of fresh biowaste after the removal of a part of the digestate formed by the preceding cycle. The digestate remaining in the batch reactor (about 50%) is mixed with the fresh one, acting as an inoculum for the microbial growing kinetics.

The process is a dry-batchwise since the solids fraction in the reacting mass is larger than 30%. The level of moisture that ensures the microbial activity inside the reactors is maintained by feeding the leachate collected from the bottom of each reactor at the top of it. It is important to highlight that the reactors are not stirred and that heat transfer and water percolation are limiting factors for the process:

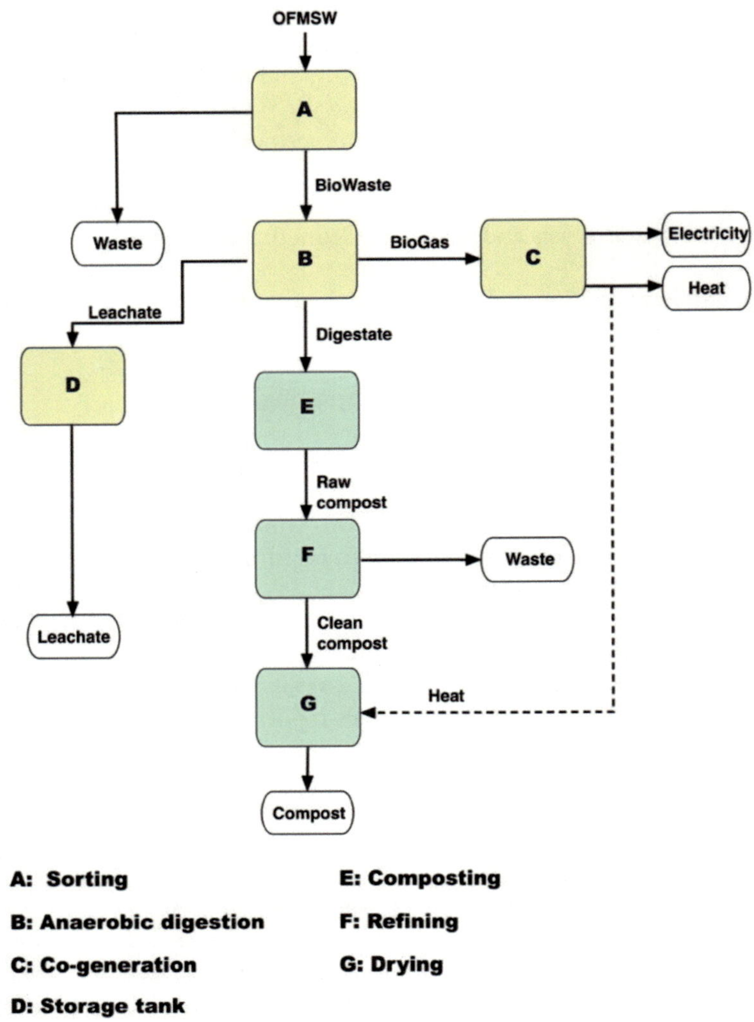

Figure 1.
Block diagram and unit processes included in the reference case.

without the recirculation of preheated leachate and the mixing with activated digestate, the process shall not occur in an appreciable way.

2.1.3 Cogeneration

The biogas generated by the anaerobic bioreactors resulted to be 5,040,000 Nm^3 per year, corresponding to a production yield of 140 Nm^3/t. This biogas is conditioned in order to remove hydrogen sulfide, ammonia, and moisture that results to be 5% in the final gas. The composition of biogas is variable, but the mean values are $60\%_v$ of methane and $40\%_v$ of carbon dioxide. The combustion of biogas is carried out in two 500 kWe Jenbacher engines that produce the electricity introduced in the public grid. The corresponding produced heat is recovered and used to enhance the composting process rate and drying the final compost.

2.1.4 Leachate storage tank

The leachate is produced during the anaerobic digestion thanks to the percolation of interstitial water from the substrate; a part of leachate is heated and

Figure 2.
Waste removed by the biowaste in the pretreatment stage.

recirculated inside the anaerobic bioreactors in order to keep the substrate humidified. The rest of the leachate is stored in a tank having a volume of 1000 m^3 and sent to the external facility to be treated and disposed. The leachate corresponds to about 30–40% of the initial biowaste.

2.1.5 Composting

The aerobic stabilization of the unconverted volatile solids occurs in order to mineralize the substrate for a period of 90 days. The composting process requires air not only for chemical oxidation of volatile solids but also for the heat removal and odor dilution in the indoor environment. The aerobic treatment requires 16,000 Nm^3 of air each Mg of digestate; that means that, in this specific case, 29,000 Nm^3/h need to be continuously extracted from the warehouse and sent to the air treatment modules in order to be cleaned up. The air treatment system receives this stream containing odorigenous molecules including organics, acids, and ammonia; it is composed of a scrubbing unit followed by two biofilters; this system is designed in order to remove odor molecules from the conveyed air stream before the diffusion in the outdoor environment. Electricity consumption for air recirculation, biofilter replacement, and wastewater treatment are expenses for this stage of the overall process.

The aerobic stabilization is followed by the maturation and refining phases (F + G). The refining process aims to remove the foreign materials and obtain a homogenized size distribution. The moisture level in the compost is lowered at 5% by using the heat recovered by section C. This phase is required in order to produce a compost that can be sold on the market of fertilizers.

2.2 The critical issues of the present configuration

The mass balance of the plant in the present configuration is reported in **Table 1**. Data are in agreement with those obtained by other anaerobic facilities assessed in the scientific and technical literature [16–18].

	Input	Intermediate	Output			
	Biowaste	Digestate	Waste	Leachate	Biogas	Compost
Mass rate, Mg/day	35,000	15,600	4200	13,278	12.34	4471
Specific indicator, Mg/Mg	—	0.45	0.12	0.38	0.13	0.13

Table 1.
Mass balance of the reference facility.

The amount of produced waste, whose amount is depending on the separate collection performance, and that of leachate both represent a negative feature as well as for environmental and economic reasons: the delivery and disposal of leachate at external facilities requires about 50 €/Mg, while the tipping fee of the waste is more than 150 €/Mg. The impact of transportation should also be included in the environmental impact assessment and in the cost evaluation since the distance between treatment plant for waste and leachate can be quite large.

Moreover, the present configuration is economically sustainable only if electricity and/or biomethane is sustained by incentives. The value of green certificates for biogas is variable, but for 1 MW biogas facilities, an indicative value of 104 €/MWh can be used [19].

Despite the incentives for the obtained products (electricity/methane), and considering that the compost has a very low value, the cost of waste disposal and that of composting (aeration) result in the increase of biowaste tipping fee that, in Italy, leads the cost of the management of separate collected waste over 190 €/t [20]. Other countries in Europe have lower tipping fees for biowaste treatment due to less restrictions about digestate use (e.g., it is not mandatory to mineralize the digestate prior to the soil scattering) and a lower cost of waste disposal due to an efficient network of waste treatment facilities. This latter cost is anyway continuously increased in the last years due to the difficulty to process inside the Europe borders the plastic waste [12].

Based on these considerations and on the European guidelines about the proximity criterion, an improvement of the industrial layout of the facility can be proposed in order to reduce the impact and the cost of the whole system. This improvement is based on the integration inside the facility boundary of the processes that allow to:

a. Treat the digestate in a more efficient way.

b. Exploit the not biodegradable waste by avoiding disposing it outside the facility boundary.

c. Treat the wastewater by using the heat produced by the integrated processes.

3. Configuration of the integrated anaerobic digestion facility

3.1 Description of the unit processes of the alternative configuration

The configuration of anaerobic digestion facility presented in **Figure 1** has been integrated with unit processes allowing:

- A sustainable production of a stable product (hydrochar) useful for agriculture purposes in place of compost

- The energy and heat production from the waste by minimizing the amount of waste to be disposed outside the facility boundary and its economic cost

- The treatment of leachate by using an effective evaporation method carried out by using the accumulated sensible heat and the recovered heat from the waste conversion

The unit processes able to reach the abovementioned targets have been reported in the integrated block diagram in **Figure 3** and described in detail in the following paragraphs.

3.1.1 Pulping and filtering

The digestate contains about 70% of moisture, as resulted from the mean values of the proximate and ultimate analyses reported in **Table 2** [21].

In this stage, it is conditioned in order to be suitable to undergo a thermal process aimed to hydrochar production, instead of the aerobic biological process aimed to compost production.

The hydrochar production process (F) occurs with very good efficiency if the organic fraction of digestate is in a pulping state, well mixed with water, and with a given water/dry matter ratio; this means that the mixing and pulping preprocessing has to be realized.

The pulping process allows to deal with another negative feature of digestate composition: the presence of foreign matter (inorganic or not reactive organics). In fact, a critical characteristic of digestate is the content of foreign matter such as glass, stones, and plastics that cannot be removed by the sorting made in the pretreatment stage where only the coarser fraction of foreign matter having a mean size larger than 5 mm is removed during the presorting; the remaining amount is not negligible and can be responsible of a dramatic decrease of the value of the final product if not removed. The pulping process between the digestate and the added water allows to separate the foreign matter by filtering the mixture following standard techniques applied in well-known wet anaerobic processes pretreatment. In this specific case, the digestate is mixed with the leachate produced by the anaerobic digestion in such a way to prepare a homogeneous pulp by respecting a given value of the parameter R (Eq. 1). In Eq. 1, $w_{digestate}$ is the weight of the digestate and $x_{H_2O,digestate}$ is the mass fraction of the moisture in the digestate.

$$R = \left(w_{digestate} \bullet x_{H_2O,digestate} + w_{H_2O,added}\right) / \left[w_{digestate} \bullet \left(1 - x_{H_2O,digestate}\right)\right] \qquad (1)$$

The pulping process can be favored by the preheating of leachate at about 70–80°C and by an intense shredding of the pulp itself. In this way, the foreign matter can be removed with very high efficacy (**Figure 4**, right) by floating (low-density fraction), by sedimentation (high-density fraction), and by sieving. The pulping creates the best conditions for this filtering/cleaning process and for the following reacting process reported in step F.

3.1.2 Hydrothermal carbonization

The hydrothermal carbonization (HTC) is a thermochemical process occurring in the presence of subcritical, liquid water: the target of this "wet or hydrous pyrolysis" is making products with higher carbon content, that are biologically inert, and with physical characteristics that make them suitable for agricultural or

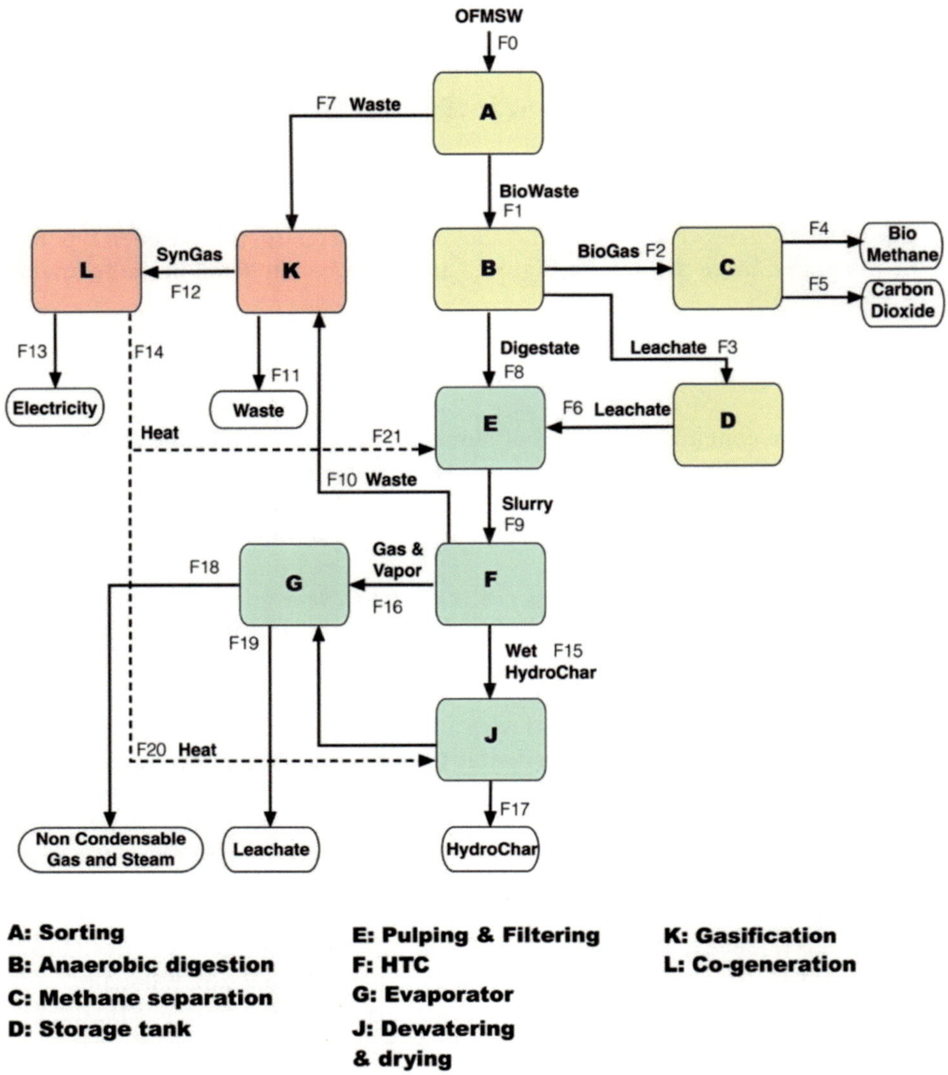

Figure 3.
Block diagram and unit processes of the proposed integrated facility.

Carbon, %	Hydrogen, %	Nitrogen, %	Oxygen, %	Ash, %	Moisture, %
32.82 ± 1.2%	4.25 ± 0.2%	2.11 ± 0.08%	33.93 ± 3.2%	26.89 ± 3.6%	69.85 ± 2.4%

Table 2.
Proximate and ultimate analyses of digestate (dry basis).

industrial purposes. The product obtained from the hydrous pyrolysis is called hydrochar to distinguish it from the biochar that is obtained by dry pyrolysis [22, 23]. During the hydrothermal process, the volatile solids contained in the digestate are surrounded by water which is kept at liquid state by allowing the pressure to rise until the endogenous value reached at the reactor temperature. The production of gases is very limited (1–5%), while dissolution of elements in the water can have a certain extent such as 5–20% [24]. The carbonization requires a reaction time of hours (1–12 h), depending on the reaction's temperature utilized, and occurs in a closed reactor by allowing the elimination of diffuse release of odors.

Figure 4.
Photos of digestate (left) and pulp (right).

Moreover, the high temperature destroys pathogens and active organic molecules. The resulted hydrochar may contribute to climate change mitigation and soil amelioration [24]. The HTC process is basically a decomposition process where chemical reactions such as hydrolysis, dehydration, decarboxylation, aromatization, and condensation occur [25]. The hydrolysis's activation energy is lower than the other reactions; this lowers the decomposition temperature of the main constituents of biomasses: cellulose and lignin, for example, decompose between 180 and 220°C.

The reactor where the HTC is carried out is a batch reactor, filled with the pulp obtained by mixing, stirring, and filtering digestate and leachate and heated up to the reaction temperature in a given heating time. A schematic process flow diagram of the HTC section is reported in **Figure 5**.

The pressure inside the reactor is determined by the temperature setup for the reaction; the temperature vs. pressure correlation can be obtained by the Antoine equation.

In the specific case of this simulation, the HTC operating conditions are reported in **Table 3**.

3.1.3 Evaporator

This stage realizes the evaporation of water under the form of steam by using the pressure difference between the reactor and the evaporator and the sensible heat of the liquid water. The evaporation allows to realize the separation of the pure water under the form of vapor from a concentrated flow, having a higher boiling temperature. The concentrated flow is rich in nitrogen and carbon; it is sterile and could be used as fertilizer, in case the regulations allow this application. The worst case is to consider it as leachate, as in the case of **Figure 3**. The steam can be condensed or lost in the atmosphere, depending on specific cases.

3.1.4 Dewatering and drying

The wet hydrochar is sent to dewatering, which is quite efficient due to its hydrophobia [26], and finally dried.

3.1.5 Gasification

The waste produced by the anaerobic digestion facility consists mainly of plastic bags and dishes, foils, and a limited amount of metals. It is basically a combustible

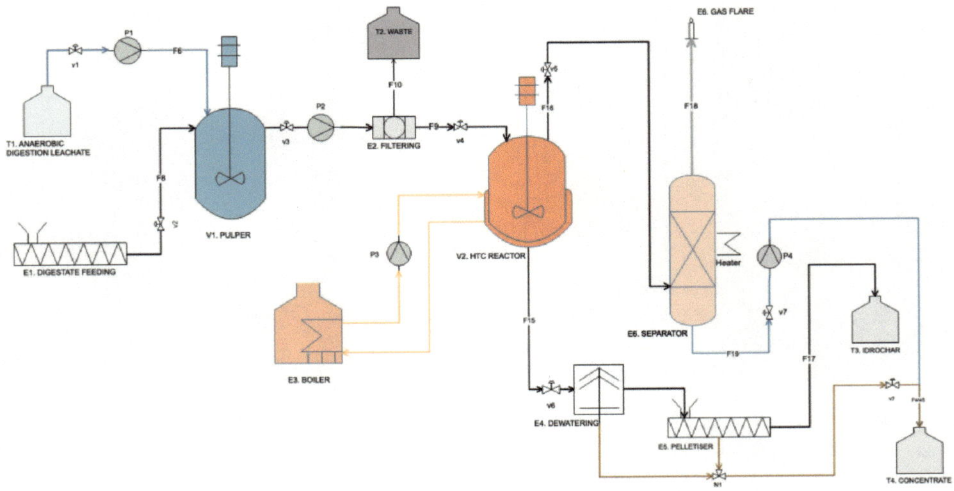

Figure 5.
Process flow diagram of HTC section.

Temperature of reaction	**°C**	**220**
Pressure	Bar	19
Water/digestate ratio	Mg/Mg	0.90
Reaction time	h	6.0
Reactor heating time	h	0.4
Reactor charge-discharge time	h	1.0
Total treatment time	h	7.4

Table 3.
HTC reactor and process parameters.

material having a moisture content until 20% and it is very dirty. Its fate is the landfilling or energy recovery in large incineration plants.

The integration of the AD plant with a small-scale gasifier allows to reduce the disposal costs and the production of electricity and heat necessary for the other processes.

The gasification unit basically consists of a downdraft reactor equipped with a fixed bed as support for the primary reactions of the combustible material with water and air, three plasma torches aligned with the bed surface, a secondary zone where homogeneous reactions take place, and a secondary stream of air that is added to favor the reactions' completion. The gasification reactions occurring on the plastic-based waste are responsible of a large production of hydrocarbons and aromatics (PAHs), partly transformed into tar, that need to be minimized in the syngas than the primary and secondary methods [13, 15, 27, 28]. In this specific case, the minimization of tar content is guaranteed by a mechanism of thermal cracking coupled with the saturation of produced radicals promoted in a secondary cracking reactor, located at syngas exit. The thermal cracking is realized by obligating the syngas by passing throughout a plasma plume composed of oxygen and hydroxyl radicals.

3.1.6 Energy production

The syngas is then sent to a cogeneration system to produce electricity and heat. Data related to the production of energy are reported in the following paragraphs.

A possible alternative is to convert the syngas into heat and use it to feed the leachate treatment process and increase the evaporation yield.

3.2 Assessment of the unit processes of the alternative configuration

3.2.1 HTC mass and energy balance

With reference to the section of the plant reported in **Figure 3** and destined to the hydrochar production and thickening of leachate, the mass balance is reported in **Table 4**. The flow ID are referred to in **Figure 3**.

The yield of hydrochar referred to the initial biowaste results to be 7%. The amount is quite similar to the compost yield, obtained by using the aerobic stabilization instead of the HTC process. The main differences are the following:

a. The HTC process requires three cycles by day, each as long as 6 hours, and two closed batch reactors to reach the given yield.

b. The HTC process does not emit odors and fugitive/diffuse emissions.

c. The HTC process requires an area quite lower than that necessary for composting.

d. The HTC can be conveniently coupled to a leachate treatment plant based on the evaporation/thickening standard processes available in the industrial market by using the heat content of liquid/vapor at the reactor outlet.

The energy balance of the HTC section is proposed in **Table 5**.

The heat necessary for the heating of the slurry from the input temperature (exit of filter) up to 220°C is 866 MJ/Mg (at 42.6 bar); once the reaction temperature is reached, the carbonization begins by absorbing heat from the environment until exothermic reactions begin. The heating time has been fixed in 0.4 h (1440 sec), so an installed heating power of about 10 MW is necessary to provide the heat in the specified time interval.

The thermal energy necessary to provide for the evaporation of water has been obtained by subtracting that requested to bring the water into vapor phase at 100°C and 1 bar to the thermal energy of the water medium present in the reactor. The amount of heat to carry out this process is 1366 MJ/Mg that corresponds to a thermal power of 9.7 MW, by assuming an evaporation time equal to that necessary for slurry's heating. By using the energy content of water after the reactions are completed, it is possible to obtain the evaporation for about 78% of water

Flow ID	From	To	Mg/year	Description
F8	B	E	15,600	Digestate
F6	D	E	14,278	Leachate
F9	E	F	29,878	Slurry
F17	J	out	2526	Hydrochar (4.3% impurities)
F18	G	out	19,437	Gas and vapor
F19	G	out	6281	Leachate (concentrate)

Table 4.
HTC mass balance.

	F (slurry heating)		F (slurry reaction)	Evaporation	
From	**E**	**F**	**F**	**F**	**F**
To	F	F	F	J	F
ID	F9	F9		F18	F18
Mass amount (for each cycle and reactor)	16.06	16.06	16.06	13.01 (liquid)	10.19 (steam)
Physical state	l	l	l	l	g
P, bar	1.5	42.6	42.6	42.6	1.0
T, °C	43.8	220.0	220.0	220.0	105.0
Flow enthalpy, MJ	1242	15,162	1213	11,661	25,581
Reaction enthalpy, MJ			−1886		
Enthalpy to be provided, MJ	13,920		Negligible	13,920	

Table 5.
Evaluation of enthalpy flows for the unit processes E and F.

(10.19/13.01); the remaining fraction remains in the liquid form by forming a thick leachate with solute.

The hydrochar can be separated by water, by centrifugation, or by other standard dewatering systems.

The most important feature is that the leachate produced by the facility is reduced at 44% of that produced by anaerobic digestion sector, without increase of thermal heat, but that used for HTC reaction.

Further evaporation is technically possible and can be also economically feasible if heat demand is fulfilled by the third section of the plant: the gasification with energy recovery.

The overall feedstock energy balance is reported in **Table 6**.

The energy balance shows in brief that:

- The biogas contains 42% of the initial feedstock energy content
- The removed waste contains 37% of the initial feedstock energy content
- The hydrochar contains 20% of the initial feedstock energy content

The energy content of waste corresponds to about 3800 MW of chemical energy that can be used to produce energy by means of a gasification process, described in the following paragraph.

3.2.2 Gasification mass and energy balance

With reference to the section of the plant reported in **Figure 3** and destined to the conversion of waste into energy and heat, the data in **Table 7** are the basis for calculation. Data refer to the the typical waste resulting from the sorting of biowaste treated in the reference facility, just before being fed to digestors. These data are the starting point for calculation of calorific value, stoichiometric oxygen demand, bottom ash production rate, and other process parameters.

	Mass flow, Mg/year	Energy content, HHV, MJ/Mg	Energy flow, MJ/day
Raw biowaste	36,000.0	7437	892,392
Waste	4320.0	18,658	268,673
Biowaste	31,680.0	5906	623,719
Of which biodegradable	30,600.0	5456	556,551
Of which foreign matter	1080.0	18,658	67,168
TOTAL IN (AD)	*31,680.0*		
Biogas	6122.4	18,380	375,100
Leachate	14,277.6		
Digestate	11,280.0	6612	248,619
Of which foreign fraction in digestate (nonbiodegradable)	1080.0	18,658	67,168
TOTAL IN (pulper)	*28,218.2*		
Digestate	11,280.0	6612	248,619
Added water/leachate	16,938.2		
Removed foreign matter	972.0	18,658	60,451
Mixture	27,246.2	2072	188,168
Of which water	24,044.6		
Of which organic	3093.6	17,596	181,451
Of which foreign matter	108.0	18,658	6717
TOTAL IN (HTC)	*27,246.2*		
Gas	624.2		
Liquid	24,096.4		
Hydrochar (dry)	2525.6	20,766	174,817
Of which foreign fraction	108.0	18,658	6717
TOTAL OUT (HTC)	*27,246.2*		

Table 6.
Energy balance.

The gasification process is carried out in a plant that is composed of two sections:

a. Thermal conversion under partial oxidation conditions

b. Energy recovery (with two options: electricity and heat or only heat)

The conversion section is composed of the gasifier itself and the thermal cracking reactor where the tar undergoes cracking and upgrading. The gasifier is fed with the waste once shredded and compacted in order to increase the bulk density at a suitable value to guarantee a constant mechanical feeding. The waste is then converted into syngas by recurring to a thermal cracking at high temperature, sustained by plasma torches, providing a thermal load of 125 kW.

The gasifier temperature as obtained by the energy balance results to be 1350°C in the homogeneous phase. The air-to-waste ratio has been fixed at 2.13 and corresponds to an equivalence ratio of 0.25.

The main input and output parameters are reported in **Table 8**.

Commodity item, #	Weight fraction, %	Ultimate analysis	Weight fraction, %
Metals	0.5%	C	51.71%
Biowaste	15%	H	9.25%
Plastics	68%	O	12.41%
Paper and cardboard	10%	N	0.85%
Glass and inerts	2%	Cl	2.42%
Wood	5%	S	0.28%
		Moisture	17.06%
		Ash	5.56%

Table 7.
Waste characterization.

Waste flow rate	0.668	Mg/h
Air flow rate	1425	Nm^3/h
Bottom ash	34.15	kg/h
Syngas flow rate (humid)	2059	kg/h
Syngas yield (dry)	1832	Nm^3/t
Syngas calorific value (dry)	5.71	MJ/Nm^3
Engine electricity generated power	832.2	kWe
Cogenerated heat	1058	kWe
Plasma torch absorption	75	kWe
Plasma torch installed power	150	kWe
Cold gas efficiency (CGE)	0.70	—

Table 8.
Gasification parameters and calculated data.

4. Conclusions

The integration between the anaerobic digestion plant of biowaste with thermochemical processes such as hydrothermal carbonization coupled with the evaporation/thickening process and gasification of nonbiodegradable waste has been proposed and described.

The mass and energy balances have been reported in order to evaluate the feasibility of the proposed integrated “smart” facility.

The advantages obtained by using the smart facility design can be briefly summarized as follows:

- The waste production destined to disposal decreased by 95%.
- The leachate to be disposed decreased by more than 66%.
- The electricity produced is 5370 MWh/year.
- The hydrochar can be sold in the market or used as fuel in the gasifier.

By an economic point of view, the capital cost of the smart facility increases by a factor of 3. Anyway, the operating costs strongly decreased by considering the savings connected to the above listed advantages.

The strong reduction of transportation of waste and leachate by trucks, the minimization of air pollution and odor emission, the possibility to cover the electricity cost by self-producing itself without grid losses, the reduction of waste volume of more 95%, and the production of clean biofuel (biomethane) and a high-added value char (hydrochar), all these positive features allow to consider the smart facility and even a green facility.

Acknowledgements

V:ALERE 2019 grant support from Università degli Studi della Campania "L. Vanvitelli" of CHIMERA project is gratefully acknowledged for financial support of the experimental activity.

Bell Production SpA and CEA SpA are acknowledged to have provided the data related to their owned technologies and facilities.

Conflict of interest

The author declares no conflict of interest.

Author details

Maria Laura Mastellone
University "Luigi Vanvitelli," DiSTABiF, Caserta, Italy

*Address all correspondence to: marialaura.mastellone@unicampania.it

References

[1] European Commission. Communication from the Commission to the Council and the European Parliament on Future Steps in Bio-Waste Management in the European Union, COM/2010/0235; 2010

[2] Zaccariello L, Cremiato R, Mastellone ML. Evaluation of municipal solid waste management performance by material flow analysis: Theoretical approach and case study. Waste Management and Research. 2015;**33**(10):871-885

[3] Schüch A, Morscheck G, A. Lemke MN. Bio-waste recycling in Germany – Further challenges. Procedia Environmental Sciences. 2016;**35**: 308-318

[4] Dahlén L, Vukicevic S, J.E. Meijer AL. Comparison of different collection systems for sorted household waste in Sweden. Waste Management. 2007;**27**:1298-1305

[5] CEA. Economic Data Related a Full-Scale Anaerobic-Aerobic Digestion Plant [Internet]. 2020. Available from: www.cea.na.it

[6] IUPAC. Compendium of Chemical Terminology Gold Book; 2014. DOI: 10.1351/goldbook. ISBN: 0-9678550-9-8

[7] Fagbemi L, Khezami L, Capart R. Pyrolysis products from different biomasses. Applied Energy [Internet]. 2001;**69**(4):293-306 Available from: https://linkinghub.elsevier.com/retrieve/pii/S0306261901000137

[8] United States Department of Agriculture. Glossary of Agricultural Terms. 2014. Available from: https://agclass.nal.usda.gov/agt.shtml

[9] Lehmann J, Joseph S. Biochar for environmental management - an introduction. In: Biochar for Environmental Management: Science and Technology. London: Earthscan; 2009. pp. 1-12

[10] Libra JA, Ro KS, Kammann C, Funke A, Berge ND, Neubauer Y, et al. Hydrothermal carbonization of biomass residuals: A comparative review of the chemistry, processes and applications of wet and dry pyrolysis. Biofuels [Internet]. 2011;**2**(1):71-106. Available from: https://www.tandfonline.com/doi/full/10.4155/bfs.10.81

[11] Mumme J, Eckervogt L, Pielert J, Diakité M, Rupp F, Kern J. Hydrothermal carbonization of anaerobically digested maize silage. Bioresource Technology [Internet]. 2011;**102**(19):9255-9260 Available from: https://linkinghub.elsevier.com/retrieve/pii/S0960852411009266

[12] Mastellone ML. Technical description and performance evaluation of different packaging plastic waste management's systems in a circular economy perspective. Science of the Total Environment [Internet]. 2020;**718**: 137233 Available from: https://linkinghub.elsevier.com/retrieve/pii/S0048969720307439

[13] Mastellone ML. Waste management and clean energy production from municipal solid waste. New York: Nova Publishers; 2015. ISBN: 978-1-63463-827-2

[14] Fernández-González JM, Grindlay AL, Serrano-Bernardo F, Rodríguez-Rojas MI, Zamorano M. Economic and environmental review of waste-to-energy systems for municipal solid waste management in medium and small municipalities. Waste Management [Internet]. 2017;**67**: 360-374 Available from: http://www.sciencedirect.com/science/article/pii/S0956053X17303057

[15] Lopez G, Artetxe M, Amutio M, Alvarez J, Bilbao J, Olazar M. Recent advances in the gasification of waste plastics. A critical overview. Renewable and Sustainable Energy Reviews [Internet]. 2018;**82**:576-596. Available from: http://www.sciencedirect.com/science/article/pii/S1364032117312832

[16] Banks CJ, Chesshire M, Heaven S, Arnold R. Anaerobic digestion of source-segregated domestic food waste: Performance assessment by mass and energy balance. Bioresource Technology [Internet]. 2011;**102**(2):612-620 Available from: https://linkinghub.elsevier.com/retrieve/pii/S0960852410013404

[17] Møller J, Christensen TH, Jansen J, La C. Anaerobic digestion: Mass balances and products. In: Solid Waste Technology & Management [Internet]. Chichester, UK: John Wiley & Sons, Ltd; 2010. pp. 618-627. Available from: http://doi.wiley.com/10.1002/9780470666883.ch39

[18] Nallathambi GV. Anaerobic digestion of biomass for methane production: A review. Biomass and Bioenergy [Internet]. 1997 Jan;**13**(1–2): 83-114 Available from: https://linkinghub.elsevier.com/retrieve/pii/S0961953497000202

[19] GSE. Economic Incentives for Biogas Electricity Production. 2020. Available from: www.gse.it

[20] ISPRA. Waste Report (Rapporto Rifiuti). 2019. Available from: http://www.isprambiente.gov.it/it

[21] Mastellone ML. Private Communication. 2020

[22] Funke A, Ziegler F. Hydrothermal carbonization of biomass: A summary and discussion of chemical mechanisms for process engineering. Biofuels, Bioproducts and Biorefining [Internet]. 2010;**4**(2):160-177. Available from: http://doi.wiley.com/10.1002/bbb.198

[23] Hu B, Wang K. WL et al. engineering carbon materials from the hydrothermal carbonization process of biomass. Advanced Materials. 2010; **22**(7):813-828

[24] Libra JA. Hydrothermal carbonization of biomass residuals: A comparative review of the chemistry, processes and applications of wet and dry pyrolysis. Biofuels. 2011;**2**(1):89-124

[25] Peterson AA, Vogel F, Lachance RP, Froeling M, Antal MJ. Thermochemical biofuel production in hydrothermal media: A review of sub- and supercritical water technologies. Energy and Envinronmental Sciences. 2008;**1**: 32-65

[26] Sharma R, Jasrotia K, Singh N, Ghosh P, Srivastava S, Sharma NR, et al. A comprehensive review on hydrothermal carbonization of biomass and its applications. Chemistry Africa [Internet]. 2020;**3**(1):1-19 Available from: http://link.springer.com/10.1007/s42250-019-00098-3

[27] Arena U, Zaccariello L, Mastellone ML. Gasification of a plastic waste in a fluidized bed of olivine. In: CFB 2008—Proceedings of the 9th International Conference on Circulating Fluidized Beds, in Conjunction with the 4th International VGB Workshop "Operating Experience with Fluidized Bed Firing Systems." 2008

[28] Hatayama A. Gasification of waste plastics by steam reforming in a fluidized bed. Journal of Material Cycles and Waste Management. 2009;**11**:144